Achieving QTS

Primary Science: Audit and Test

Fourth edition

Jenny Byrne
Andri Christodoulou
John Sharp

D1350724

Los Angeles | London | New Delhi
Singapore | Washington DC

Learning Matters
An imprint of SAGE Publications Ltd
1 Oliver's Yard
55 City Road
London EC1Y 1SP

SAGE Publications Inc.
2455 Teller Road
Thousand Oaks, California 91320

SAGE Publications India Pvt Ltd
B 1/I 1 Mohan Cooperative Industrial Area
Mathura Road
New Delhi 110 044

SAGE Publications Asia-Pacific Pte Ltd
3 Church Street
#10-04 Samsung Hub
Singapore 049483

Editor: Amy Thornton
Production controller: Chris Marke
Project management: Deer Park Productions, Tavistock, Devon
Marketing manager: Catherine Slinn
Cover design: Wendy Scott
Typeset by: C&M Digitals (P) Ltd, Chennai, India
Printed by CPI Goup (UK) Ltd, Croydon, CR0 4YY

First published in 2001 by Learning Matters Ltd

Reprinted in 2002. Second edition published in 2003. Reprinted in 2003, 2004, 2005, 2006.

Third edition published in 2007. Reprinted in 2007 (twice), 2008, 2009, 2011.
Fourth edition published in 2014.

Library of Congress Control Number: 2013953925

British Library Cataloguing in Publication Data

A catalogue record for this book is available from the British Library

ISBN 978-1-4462-8273-1 (pbk)
ISBN 978-1-4462-8272-4

Contents

Introduction

About this book

This book has been written to support the subject knowledge learning of all primary trainee teachers on courses of Initial Teacher Training (ITT) in England and other parts of the UK where a secure subject knowledge and understanding of science is required for the award of Qualified Teacher Status (QTS). A secure subject knowledge and understanding of science is now widely acknowledged as a critical factor at every point in the complex process of planning, teaching and assessing science. The audit and test materials presented here will help you to identify your own strengths and weaknesses in science. As you revise, you can revisit these to help you monitor and evaluate your own progress towards QTS.

Part 1: Science background;

Part 2: Interest in science;

Part 3: Perceived competence and confidence in science;

Part 4: Science test;

Part 5: Answers to test questions including children's common misconceptions of science concepts;

Part 6: Targets for further development;

Part 7: Revision and further reading.

It is quite likely that you will be required to undertake further auditing and testing of your subject knowledge and understanding of science at the start of your own course of ITT. You may wish to retain the audit and test result details for your own records and use them to return to as and when necessary. Your ITT provider may also wish to use them for their own records.

While you may indeed find the process of auditing and testing your science subject knowledge particularly daunting, especially if you were last taught or thought about science several years ago, most people simply take it all in their stride and you should aim to do the same. Auditing and testing are simply two forms of assessment, and you will be assessed in many different ways throughout your career in teaching, not just when you train. There is certainly nothing to worry about when auditing and testing yourself in the comfort of your own home. Your ITT provider will also take every step they can to help minimise the apparent stress of auditing and testing as they guide you towards your goal of becoming an effective and successful primary school teacher.

For trainees wishing to undertake some revision or who feel the need for a science study aid, there are several excellent books written specifically for primary trainees with diverse backgrounds in science. Some of these are listed in Part 7. All are available from good booksellers. We, of course, would recommend *Primary Science: Knowledge and Understanding* (6th edition) by Peacock et al. (2012) from the *Learning Matters Achieving QTS Series*. Almost all of the test questions and diagrams here are drawn from this source. Similarly, we would also recommend *Primary Science: Teaching Theory and Practice* (6th edition) by Sharp et al. (2012).

The Teachers' Standards (2012)

A statutory framework for the career-long professional development of teachers in England has been designed by the Department for Education. The Standards apply to all teachers including those working towards QTS, which trainees must meet if they are to be awarded QTS.

The Standards include aspects of professional knowledge and understanding which are a requirement in order to be able to teach effectively and ensure pupil progress. Teachers are required to demonstrate good subject and curriculum knowledge *(Teacher Standard 3)*. This means that they are required to know and understand the *National Curriculum for Science* and to have a secure knowledge and understanding of the science subject knowledge as appropriate to the age range for which they are trained. With respect to science, this will mean offering pupils enjoyable and creative learning opportunities, which may be integrated with other subject areas. A high level of science subject knowledge is an essential aspect of teaching in this manner. In addition to secure subject knowledge, teachers are required to be aware of common misconceptions in primary science and of ways in which these might be addressed in the classroom. The audit and test materials in this book include many aspects of the specific science subject knowledge from the *relevant parts* of the *National Curriculum* programmes of study at Key Stage 3, which you will need to know and understand in order to adequately plan for, teach and assess the pupils in your classroom at Key Stages 1 and 2.

In terms of the professional skills set out in the Standards, it is vital that your subject knowledge is sufficiently secure for you to feel confident when teaching and assessing children's learning. Strong subject knowledge will enable you to understand the concepts you teach so that you can explain them effectively and offer examples, and to help your pupils investigate them and develop their own understandings. Strong subject knowledge will enable you to identify specifically what your pupils can do, and what they need to learn next. It will help you devise effective questions and provide feedback to move learning on.

The Standards also require you to show that you are beginning to address your personal professional development by taking steps to identify and meet your own professional needs. This book has been designed to assist you in doing this by helping you to identify particular areas of your subject knowledge that require further study.

The latest programmes of study for Key Stage 3 that are relevant to teaching at Key Stages 1 and 2 include the following.

- Working scientifically

  o Scientific attitudes
  o Experimental skills and investigations
  o Analysis and evaluation
  o Measurement

- Subject content

  o Biology (including the structure and function of living organisms, interactions and interdependency, genetics and evolution)
  o Chemistry (including materials and their properties, the particulate nature of matter, and Earth and atmosphere)
  o Physics (including motion and forces, sound, light, electricity, magnetism and space physics)

The audit materials presented here will introduce you to all of the content items listed above in detail. You will also have the opportunity to test your knowledge and understanding in most of these areas.

Science: the statutory framework
The Primary National Curriculum in England (2014)

Schools have a statutory duty to teach the National Curriculum, which was first introduced in 1989. It is organised on the basis of four Key Stages, of which Key Stage 1 for 5–7 year olds (Years 1 and 2) and Key Stage 2 for 7–11 year olds (Years 3–6) cover the primary years. The National Curriculum for each Key Stage includes programmes of study, which set out the science that children should be taught, and non-statutory notes and guidance for each subject area. A brief summary of the primary science programmes of study is as follows.

Key Stage 1 (Years 1 and 2)

- Working scientifically
- All living things including i) plants, ii) animals, including humans, iii) habitats
- Everyday materials and their uses
- Seasonal changes

Lower Key Stage 2 (Years 3 and 4)

- Working scientifically
- All living things including i) plants, ii) animals, including humans and iii) habitats
- Everyday materials including i) rocks, and ii) states of matter
- Forces and magnets
- Electricity

Upper Key Stage 2 (Years 5 and 6)

- Working scientifically
- All living things including i) plants, ii) animals including humans, iii) habitats and iv) evolution and inheritance
- Properties and changes of materials
- Light
- Forces
- Electricity
- Earth and space

'Working scientifically' aims to develop pupils' knowledge and understanding of the nature, processes and methods of science. It includes aspects such as experimental skills, conducting investigations, handling information, problem solving, using models to describe scientific ideas, using scientific evidence to support or refute ideas, and measurement. 'Working scientifically' should not be taught as a separate component of the National Curriculum for Science. Instead, aspects of 'working scientifically' should function as the context in which the subject knowledge as outlined above for Key Stages 1 and 2, is to be taught.

The Foundation Stage

The scope of the National Curriculum was extended in 2002 to incorporate the Foundation Stage as a distinct stage of education for children aged from 3 to 5. The Early Years curriculum, set out in *Statutory Framework for the Early Years Foundation Stage* (DfE, 2012), covers four broad areas of learning, including 'Understanding the world', and involves helping children to make sense of their physical world through exploration, observation and finding out about technology and the environment. Statutory early learning goals describe what children are expected to achieve in these areas by the end of the Reception year and provide the basis for the Early Years curriculum. Children are assessed against these goals prior to entering Year 1 and are given an Early Years Foundation Stage Profile, which provides an overall picture of the child's progress against expected levels.

References

DfE (2014) *The National Curriculum in England.* London: HMSO. (Also available online at www.education.gov.uk/schools/teachingandlearning/curriculum/nationalcurriculum2014/ [accessed 23.10.2013].)

DfE(2012) *Statutory Framework for the Early Years Foundation Stage*. Cheshire: DfE. (Also available on line at www.education.gov.uk/publications/standard/AllPublications/Page1/DFE-00023-2012 [accessed 23.10.2013].)

DfE (2012) *Teachers' Standards.* London: HMSO. (Available online at www.gov.uk/government/publications/teachers-standards [accessed 23.10.2013].)

Peacock, G., Sharp, J., Johnsey, R. and Wright, D. (2012) *Primary Science: Knowledge and Understanding.* 6th ed. London: Learning Matters.

Sharp, J., Peacock, G., Johnsey, R., Simon, S., Smith, R., Cross, A. and Harris, D.(2012) *Primary Science: Teaching Theory and Practice.* 6th ed. London: Learning Matters.

Part 1: Science background

Provide as many background details as you can. Don't worry if it looks a bit 'blank' in places, you won't be alone. Unless you are a prospective science specialist, what else would you expect?

► **personal details**

Name

Date of birth

Year(s) of course

Subject specialism

Elected Key Stage

► **science qualifications**

GCSE/O level (equivalent)

Date taken

Grade(s)

GCE A level (equivalent)

Date taken

Grade(s)

► **science degree**

Subject

Year of graduation

Class of degree

Other science courses

► **other** (e.g. work related)

Part 2: Interest in science

A positive attitude towards science will help you to learn and teach it well, whether it is your favourite subject or not. Be honest with yourself and think carefully about your responses below. It is possible, for example, that you might have a healthy interest in science even if you don't think you know too much about it right now.

Circle as appropriate using the key provided.

1 = **I am very interested in science.**

2 = **I am interested in science.**

3 = **I am uncertain about my interest in science.**

4 = **I am not interested in science.**

Interest   **1**   **2**   **3**   **4**

A 1 or a 2 is fantastic, a 3 encouraging, a 4 – well, science is not everybody's thing. Reflect critically on your attitude towards science, positive or negative, and use the space below to comment further. Can you identify the experiences which gave rise to your interest or lack of it?

experiences statement

Part 3: Perceived competence and confidence in science

It is entirely possible that as you respond to the following sections you might notice that you feel quite competent in an area of science, or even a 'strand' within it, but lack the confidence to teach it. Competence and confidence are clearly quite different things. By the end of your training you will feel better about both.

Competence

There are rather a lot of areas within the self-audit and you will need some time to read through and complete this part thoroughly. The 'strands' are reproduced here to introduce you to the requirements. You do not need to know about or feel competent with everything listed here right now. There will be plenty of time for this later.

Please respond to the following statements using the key provided.

1 = Very good. Existing competence perceived as *exceeding* the requirements.

2 = Good. Existing competence perceived as *meeting* the requirements *comfortably*.

3 = Adequate. Existing competence perceived as *meeting* the requirements but *some* uncertainties still exist.

4 = Not good. Existing competence perceived as *not meeting* the requirements.

Working scientifically

To underpin and to support the effective teaching of primary science, you should know and understand something of the nature of scientific knowledge and practices, for example:

a) Scientific attitudes and views of the nature of science	1	2	3	4
• science is a way of making sense of natural phenomena	1	2	3	4
• scientific knowledge may change as existing evidence or observations are reinterpreted by scientists based on new ideas and developments	1	2	3	4
• scientific knowledge and explanations may change as new evidence is collected and thinking is challenged	1	2	3	4
• science is a collaborative activity which involves a community of scientists and others developing more powerful ways of understanding the natural world	1	2	3	4

- science does not explain every phenomenon 1 2 3 4
- scientific knowledge and understanding can be used in solving a range of problems but the available scientific evidence is often limited, and its application to everyday problems often entails consideration of ethical or moral questions 1 2 3 4
- science has played a part in many of the things that you use 1 2 3 4

b) Experimental skills and investigations **1 2 3 4**

- the fact that not all questions can be investigated practically 1 2 3 4
- how to construct questions that can be investigated, including considering the distinction between a guess, a prediction and a hypothesis 1 2 3 4
- how to plan investigations appropriate to the question asked, the resources available and the context in which they are carried out 1 2 3 4
- the structure and use of controlled experiments, taking into account all the relevant variables to allow valid comparison of different sets of data 1 2 3 4
- the ways in which sample size can be selected, how this will influence the outcomes of an investigation and how this can be recognised when findings are interpreted 1 2 3 4
- making informed decisions about which types of scientific enquiry are the most appropriate for the collection, analysis and interpretation of a set of evidence that will answer the original question 1 2 3 4
- the importance of selecting and using equipment correctly in order to gather evidence at the required level of detail 1 2 3 4
- the fact that outcomes of an investigation should be considered and evaluated in the light of the original question and the wider body of available and relevant scientific evidence 1 2 3 4

c) Analysis and evaluation **1 2 3 4**

- the different ways in which objects and organisms can be identified and classified, including the construction and use of keys 1 2 3 4
- the need to record the relevant evidence accurately and, where appropriate at suitable time intervals, using appropriate techniques, including tables, histograms, graphs or electronic devices 1 2 3 4

- the nature of variables, including identification of categoric, independent and dependent variables, and recognition of discrete and continuous variables $\qquad$ 1 2 3 4

- possible reasons for experimental findings not supporting accepted scientific theories and evidence, including: extent of available evidence, natural variation in measurements, limitations in resources and experimental design $\qquad$ 1 2 3 4

- the need for conclusions drawn to be based on reasoned argument, using evidence and making links to predictions and hypotheses $\qquad$ 1 2 3 4

- the need for accuracy and precision in observations of measurements, in the replication of readings, in the control of variables and in the acknowledgement of sources of evidence in order to establish the reproducibility, reliability and validity of evidence $\qquad$ 1 2 3 4

- the importance of peer review and publishing results in the process of evaluating and establishing new scientific knowledge $\qquad$ 1 2 3 4

d) Measurement   1   2   3   4

- the appropriate units, including SI units, which should be used to quantify the different types of measurements required $\qquad$ 1 2 3 4

- the variety of ways to collect evidence including techniques for observing, measuring, testing and controlling variables, carrying out surveys, sampling, using models and interrogating secondary sources $\qquad$ 1 2 3 4

- the different ways in which evidence can be analysed such as looking for patterns and trends using simple mathematical devices (e.g. means, scattergrams, line graphs), recognising that the form of analysis chosen should be matched to the type of evidence available $\qquad$ 1 2 3 4

e) The need for clear and precise
forms of communication in science   1   2   3   4

- the correct scientific terminology for phenomena, events and processes $\qquad$ 1 2 3 4

- the accepted scientific terminology, forms of representation, symbols and conventions $\qquad$ 1 2 3 4

- a wide range of methods, including diagrams, drawings, graphs, tables and charts, to record and present information in an appropriate and systematic manner   1   2   3   4

f) Health and safety requirements and how to implement them

1   **2**   **3**   **4**

- the major legal requirements for health and safety, including restrictions on keeping living things in the classroom   1   2   3   4
- the fact that every activity involves an element of risk which should be assessed and allowed for in planning and organising it   1   2   3   4
- the accepted actions and procedures in the event of an accident   1   2   3   4

Biology

To underpin and to support the effective teaching of primary science, you should know and understand something of living organisms, their behaviour and health, for example:

a) structure and function of living organisms

1   **2**   **3**   **4**

- the differences between things that are living and things that have never been alive   1   2   3   4
- that organisms have the potential to carry out the life processes of nutrition, movement, growth, reproduction, respiration, sensitivity and excretion   1   2   3   4
- that most organisms are made up of cells and almost all cells have a nucleus which controls their activities   1   2   3   4
- that life processes of multi-cellular organisms are supported by the organisation of cells into tissues, organs and organ systems that carry out specialised functions   1   2   3   4
- that humans have senses which enable them to be aware of the world around them   1   2   3   4
- naming the main external parts of the human body   1   2   3   4
- the functions of teeth and the importance of dental care   1   2   3   4
- a simple model of the structure of the heart and how it acts as a pump   1   2   3   4
- how blood circulates in the body through arteries and veins   1   2   3   4
- the effect of exercise and rest on pulse rate   1   2   3   4
- that humans have skeletons and muscles to support their bodies and to help them move   1   2   3   4
- the main stages of the human life cycle including fertilisation, foetal development and adolescence   1   2   3   4

- the functions of nutrition, circulation, movement, growth and reproduction in humans

  1   2   3   4

- how to recognise and name the leaf, flower, stem and root of flowering plants

  1   2   3   4

- that plant growth is affected by the availability of light and water, and by temperature

  1   2   3   4

- that plants need light to produce food for growth, and the importance of the leaf in this process

  1   2   3   4

- that the root anchors the plant, and that water and nutrients are taken in through the root and transported through the stem to other parts of the plant

  1   2   3   4

- about the life cycle of flowering plants, including pollination, seed production, seed dispersal and germination

  1   2   3   4

- that reproduction is necessary for a completed life cycle

  1   2   3   4

- that individual organisms eventually die

  1   2   3   4

- how the health of an organism can be affected by a range of factors, for example, in humans, drugs, exercise and other physical, mental and environmental factors

  1   2   3   4

- that health can be threatened by a variety of agents

  1   2   3   4

- that organisms have various ways of keeping themselves healthy

  1   2   3   4

- that food is needed for activity and for growth, and that an adequate, varied and balanced diet is needed to keep healthy

  1   2   3   4

b) interactions and interdependencies

1   2   3   4

- that a diversity of organisms is found in most habitats

  1   2   3   4

- that the organisms, including humans, in an ecosystem interact with each other and with the physical aspects of the environment

  1   2   3   4

- that the behaviour of organisms is influenced by internal and external factors and can be investigated and measured

  1   2   3   4

- that food chains show feeding relationships in an ecosystem

  1   2   3   4

- that nearly all food chains start with green plants

  1   2   3   4

- that micro-organisms are widely distributed

  1   2   3   4

- that humans affect the environment in various ways

  1   2   3   4

c) genetics and evolution

1   2   3   4

- variation exists between and within the same species of living things

  1   2   3   4

- a diversity of organisms exist, and include bacteria, fungi, plants and animals

  1   2   3   4

- living things can be classified according to observable similarities and differences and identified using keys   1   2   3   4
- the principal agent controlling the characteristics and working of cells and organisms is their genetic material, DNA   1   2   3   4
- reproduction results in DNA from the parent or parents being passed on to future generations   1   2   3   4
- variation within a species is due to genetic and environmental factors   1   2   3   4
- before reproduction, the genetic material of an organism is replicated   1   2   3   4
- mutations may occur during the process of DNA replication and during sexual reproduction, and genetic material will inevitably be recombined, both of which will cause variation in the offspring; in asexual reproduction (cloning) the amount of variation is characteristically very small and the offspring look exactly like the parent   1   2   3   4
- most biologists believe that variation caused by genetic mutation and recombination, coupled with interaction between organisms and their environment, leads to natural selection and evolutionary change   1   2   3   4
- variation can be manipulated by humans through genetic engineering and selective breeding   1   2   3   4
- a species is a group of organisms which can interbreed to produce fertile offspring   1   2   3   4

Chemistry

To underpin and to support the effective teaching of primary science, you should know and understand something of materials, their structure and their properties, for example:

a) materials and their properties including the particulate nature of matter   **1**   **2**   **3**   **4**

- the types of particles that make up all materials include atoms, protons, neutrons and electrons   1   2   3   4
- when atoms of different elements combine, the resulting material is a compound   1   2   3   4
- atoms can be held together in different ways   1   2   3   4
- the properties of a compound depend on the way in which the particles making it up are arranged and held together, such as in molecules and giant structures   1   2   3   4

- in chemical reactions new substances are formed   1   2   3   4
- physical changes involve changes in the arrangement and spacing of particles but no new substances are formed   1   2   3   4
- that some changes can be reversed and some cannot   1   2   3   4
- the properties of materials can often be predicted from a knowledge of their structures, and vice versa, but can also depend on their shape and size   1   2   3   4
- that materials are chosen for specific uses on the basis of their properties   1   2   3   4
- that some materials are better thermal insulators than others   1   2   3   4
- that some materials are better electrical conductors than others   1   2   3   4
- that mixtures of materials can be separated in a variety of different ways   1   2   3   4
- most materials can exist as a solid, liquid and gas, depending on conditions   1   2   3   4
- changes of state can be brought about by transferring energy   1   2   3   4
- finely divided substances still contain many atoms and molecules   1   2   3   4
- the movement of particles explains the properties of solids, liquids and gases and of changes such as dissolving, melting and evaporating   1   2   3   4
- during chemical changes bonds joining atoms together are broken and new bonds are formed   1   2   3   4
- mass is conserved in physical and chemical changes   1   2   3   4

b) Earth and atmosphere
    1   2   3   4

- rocks and soils can be grouped on the basis of characteristics, including appearance, texture and permeability   1   2   3   4
- rocks are formed in a variety of ways, including sedimentation and volcanic activity   1   2   3   4
- geological activity is caused by chemical and physical processes   1   2   3   4
- human activity and natural processes can lead to changes in the environment   1   2   3   4

Physics

To underpin and to support the effective teaching of primary science, you should know and understand something of physical processes, for example:

a) motion and forces

	1	**2**	**3**	**4**
• that both pushes and pulls are examples of forces	1	2	3	4
• when an object is stationary or moving at a steady speed in a straight line, the forces acting on it are balanced	1	2	3	4
• balanced forces produce no change in the movement of an object or shape of an object, whereas unbalanced forces acting on an object can change its motion (speed and/or direction) or its shape	1	2	3	4
• the change in movement and/or shape of an object depends on the magnitude and direction of the force acting on it	1	2	3	4
• forces such as (solid) friction, air resistance and water resistance oppose the relative motion between an object and what it is touching	1	2	3	4
• in most situations there are forces such as friction retarding the motion of objects and so a driving force is needed to keep them moving at a steady speed	1	2	3	4
• frictional forces between surfaces can also enable motion, e.g. by opposing the relative movement between shoe and floor or tyre and road	1	2	3	4
• that when springs and elastic bands are stretched they exert a force on whatever is stretching them	1	2	3	4
• that when springs are compressed they exert a force on whatever is compressing them	1	2	3	4
• forces are measured in newtons	1	2	3	4
• the mass of an object is the amount of matter in it; mass is measured in grams	1	2	3	4
• gravitational attraction exists between all objects; this depends on the mass of the respective objects and how far apart they are	1	2	3	4
• the weight of an object is a force caused by the gravitational attraction between the Earth and the object and directed towards the centre of the Earth	1	2	3	4
• a specific object will have the same mass on the Earth and on the Moon because it contains the same amount of matter	1	2	3	4
• an object will weigh more on the Earth than on the Moon because the Earth has greater mass and exerts greater gravitational attraction than the Moon	1	2	3	4
• objects of different mass dropped at the same instance from the same point will land at the same time unless the air resistance is different	1	2	3	4
• speed is the distance travelled by an object in a given time	1	2	3	4

- the relationship between speed, distance and time, and the distinction between speed and acceleration   1   2   3   4

b) sound   **1   2   3   4**

- there are many kinds of sound and many sources of sound   1   2   3   4
- sounds are made when objects vibrate but vibrations are not always directly visible   1   2   3   4
- sound travels through a medium away from a vibrating source   1   2   3   4
- sound travels through materials (solids, liquids and gases)   1   2   3   4
- sound waves can differ in amplitude and frequency and this leads to differences in loudness and pitch respectively   1   2   3   4
- sounds are heard when vibrations from an object enter the ears causing the eardrums to vibrate and impulses to be carried to the brain   1   2   3   4

c) light   **1   2   3   4**

- light travels from a source   1   2   3   4
- darkness is the absence of light   1   2   3   4
- light travels in a straight line unless something prevents it from doing so, e.g. reflection or scattering, and this can be used to explain the formation of shadows   1   2   3   4
- why the size of shadows might change   1   2   3   4
- the distinction between reflection and scattering and how images are formed in a mirror   1   2   3   4
- the colour of an object depends on the wavelength of light that it scatters, e.g. a black object scatters little light and absorbs light of all visual wavelengths; a green object scatters more green light than other colours which it absorbs more   1   2   3   4
- light can be broken into different colours and different colours of light can be combined to appear as a new colour   1   2   3   4
- objects are seen when light is emitted or reflected from them and enters the eye through the pupil, causing the retina to send messages, carried by nerves, to the brain   1   2   3   4

d) electricity   **1   2   3   4**

- all matter is made up of particles which include electrons – these carry a negative charge   1   2   3   4
- 'resistance' is a measure of the difficulty of flow of electrons in the material   1   2   3   4

- when a cell (or battery) is attached to a circuit, it provides a 'push' which causes electrons to move in one direction around the circuit; this movement (flow) of electrons is called current (measured in amps)   1   2   3   4
- current is not consumed and is the same in all parts of a simple series circuit   1   2   3   4
- voltage (measured in volts) is a measure of the energy per unit charge and this might be considered as driving the current   1   2   3   4
- short circuits may cause wires to heat up; fuses are electrical safety devices that are triggered by short circuits   1   2   3   4
- the power (measured in watts) of a device such as a bulb or motor is the rate at which energy is transferred to the device   1   2   3   4
- as moving electrons collide with fixed atoms in a circuit they make the atoms vibrate more; this vibration causes components such as bulb filaments to get hot and emit light   1   2   3   4
- how to represent simple series and parallel circuits using standard symbols in circuit diagrams   1   2   3   4
- how to construct simple circuits involving batteries, wires, bulbs and buzzers on the basis of drawings and diagrams   1   2   3   4

e) magnetism   **1   2   3   4**

- magnets have poles; like poles repel and unlike poles attract   1   2   3   4
- magnetism can act over a distance, so magnets can exert forces on objects with which they are not in contact   1   2   3   4
- explain how a compass works based on ideas such as the Earth's magnetic field   1   2   3   4

f) space physics   **1   2   3   4**

- the Universe includes galaxies which include stars, such as the Sun   1   2   3   4
- the Sun is one star in our galaxy and is at the centre of our Solar System   1   2   3   4
- the order of the planets in our Solar System, their major features and relative distances from the Sun, which they orbit   1   2   3   4
- that the Sun, the Earth and the Moon are approximately spherical   1   2   3   4
- the explanation of day and night and the evidence for it   1   2   3   4
- the explanations for the phases of the Moon and eclipses   1   2   3   4
- that the Moon is not a source of light   1   2   3   4
- the explanations for the seasons and length of year   1   2   3   4
- that the position of the Sun appears to change during the day, and how shadows change as this happens   1   2   3   4

Making sense of your perceived competence

Look back over your **perceived competency** grades. Summarise each area in the following table by looking at the distribution of responses. In Physical processes, for example, if you ticked lots of 2s, 3s and 4s but no 1s, you should fill in your range as 2s to 4s. If, say, you ticked more 3s than anything else, you should fill in your mode, the most frequently occurring response, as mostly 3s.

	range	mode
Working scientifically	_____	_____
Biology	_____	_____
Chemistry	_____	_____
Physics	_____	_____

Mostly 1s   Areas summarised as mostly 1s suggest that most competency requirements are exceeded. Your perceived competence would place you at a level beyond that of a non-science specialist. Well done.

Mostly 2s   Areas summarised as mostly 2s suggest that most competency requirements are met comfortably. Some attention is necessary locally, certainly in the weaker elements. Your perceived competence places you at a level about that of a non-science specialist. With this sort of profile you probably have little to worry about.

Mostly 3s   Areas summarised as mostly 3s suggest that most competency requirements are met adequately. However, attention is necessary throughout. Your perceived competence places you at a level best described as approaching that specified for a non-science specialist. You are probably in good company and with a little effort you will be up there with the best of them.

Mostly 4s   Areas summarised as mostly 4s suggest that most competency requirements are hardly being met at all. But remember, you only have to get there by the end of your training – not before! Given the nature of the requirements, a profile like this is not surprising; it would not concern us at this stage, so don't let it concern you.

Confidence

Examine the programmes of study for Key Stages 1 and 2 in the National Science Curriculum below carefully. Overall, how would you describe your confidence in terms of **teaching** them?

Please respond using the key provided.

1 = Very good. Might even feel happy to support colleagues!

2 = Good. Further professional development required in some aspects.

3 = Adequate. Further professional development required in most aspects.

4 = Not good. Further professional development essential in all aspects.

Working scientifically (Key Stages 1 and 2)

	1	2	3	4
• scientific attitudes and views of the nature of science	1	2	3	4
• experimental skills and investigations	1	2	3	4
• analysis and evaluation	1	2	3	4
• measurement	1	2	3	4
• communication and health and safety	1	2	3	4

Key Stage 1

	1	2	3	4
• plants	1	2	3	4
• animals, including humans	1	2	3	4
• all living things and their habitats	1	2	3	4
• everyday materials and their uses	1	2	3	4
• seasonal changes	1	2	3	4

Lower Key Stage 2

	1	2	3	4
• plants	1	2	3	4
• animals, including humans	1	2	3	4
• all living things and their habitat	1	2	3	4
• rocks	1	2	3	4
• states of matter	1	2	3	4
• light	1	2	3	4
• sound	1	2	3	4
• forces and magnets	1	2	3	4
• electricity	1	2	3	4

Upper Key Stage 2

	1	2	3	4
• all living things and their habitat	1	2	3	4
• animals, including humans	1	2	3	4
• evolution and inheritance	1	2	3	4
• properties and changes of materials	1	2	3	4
• Earth and space	1	2	3	4
• forces	1	2	3	4
• light	1	2	3	4
• electricity	1	2	3	4

Making sense of your perceived confidence

1s and 2s are fantastic – what has kept you away from the profession for so long! 3s are encouraging and we would imagine that many people would have this sort of profile. If you ticked any 4s, don't worry. You are being very honest with yourself and that is good. If you really felt so confident about teaching science now there would not be any point in training you to do it, would there?

Reflect critically on your perceived confidence about teaching science and use the space below to comment further. Can you identify the 'source' of your confidence or the 'source' of your lack of it?

confidence statement

Part 4: Science test

Your own perception of competence and confidence is one thing, but how would you do if actually put to the test? As always, it doesn't matter how well or how badly you test now; you will have lots of time to make up for the science you have forgotten or simply never knew in the first place. The following pages explore your knowledge and understanding in many key areas of primary science using a variety of established test techniques. Take as long as you like and try not to cheat too much by looking at the answers! The marking system is fairly straightforward and easy to use. (As it is impossible to monitor, give yourself 1 mark for every correct answer!)

Biology
Structure and function of living organisms: plants

1 Match the labels below with parts **A** to **H** shown in the diagram of the flowering plant.

leaf

root system

flower

lateral root

root hair

stem

shoot system

tap root

[8 MARKS]

Complete the following sentences by inserting the most appropriate words.

Roots _____ plants firmly in the ground. They are also responsible for the uptake of _____ and _____ from the soil. Stems hold plants upright, spread out leaves for _____ and elevate flowers for _____. Most leaves are green due to the presence of _____. Flowers are the structures of most plants that are responsible for _____.

[7 MARKS]

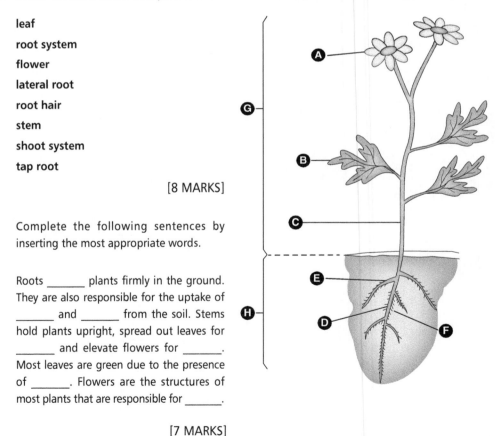

2 In the diagram of the flower section shown, label the parts **A** to **H**.

anther
sepal
receptacle
style
petal
ovary with ovules
filament
stigma

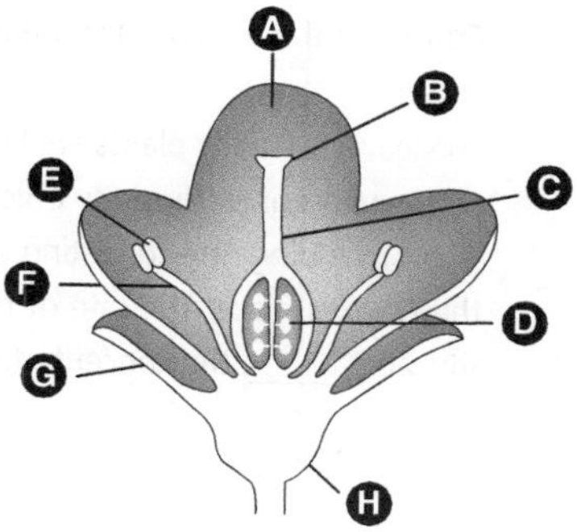

[8 MARKS]

Complete the following sentences by inserting the most appropriate words.

Flowers usually consist of five main elements (though these are not always present or immediately obvious): often brightly coloured and scented petals which attract _____, papery sepals which _____ the flower in bud, male and female reproductive organs which work together to ensure the _____ of each species, and a receptacle which _____ the weight of everything.

[4 MARKS]

3 The male reproductive organs or stamen of a flowering plant includes:

(a) petals, sepals and a receptacle
(b) the stigma, style and ovary
(c) the anther and filament

[1 MARK]

4 The female reproductive organs or carpel of a flowering plant includes:

(a) the stigma, style and ovary
(b) the anther and filament
(c) petals, sepals and a receptacle

[1 MARK]

5 In the diagram of the leaf section shown, label the parts **A** to **H**.

upper epidermal cells
spongy mesophyll cells
lower epidermal cells
waxy cuticle
guard cell
palisade mesophyll cells
'vein' (phloem and xylem vessels)
stoma

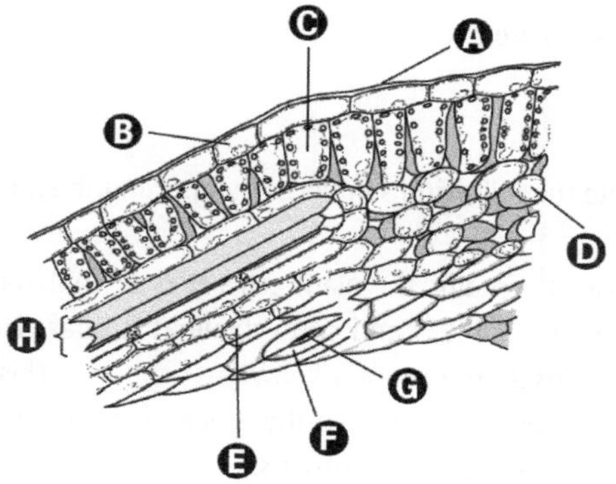

[8 MARKS]

21

Complete the following sentences by inserting the most appropriate words.

The leaves of many plants are large and _____ in order to trap sunlight and to make the process of photosynthesis particularly efficient. A cross-section through a leaf blade or _____ reveals several layers including one consisting of palisade cells within the palisade mesophyll where most photosynthesis takes place. The rate of photosynthesis is affected by several factors including _____ intensity, air _____, the concentration of _____ in the atmosphere, and _____ availability.

[6 MARKS]

6 Rearrange the following items to write out a word equation for photosynthesis.

energy in sunlight

glucose

carbon dioxide

oxygen

chlorophyll

water

[3 MARKS]

7 In the diagram of the generalised plant cell (green tissue) shown, label the parts **A** to **H**.

chloroplast with chlorophyll

cellulose cell wall

nucleus

ribosome

cell membrane

cytoplasm

mitochondrion

sap-filled vacuole

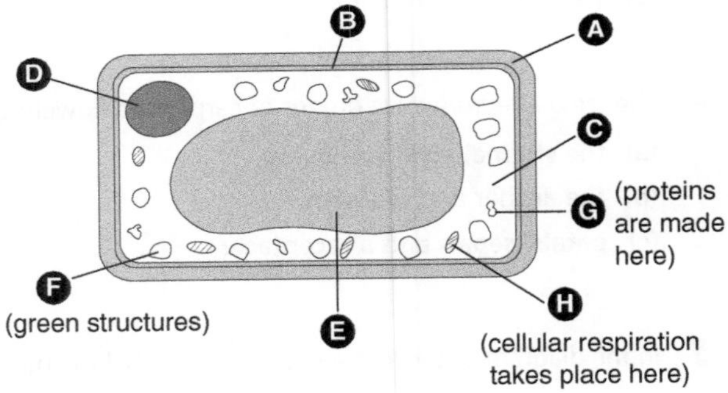

(green structures)

(proteins are made here)

(cellular respiration takes place here)

[8 MARKS]

Complete the following sentences by inserting the most appropriate words.

The nucleus contains the plant cell's genetic material or _____, which ultimately determines what type of cell it is and controls what it does. The genetic material also has the ability to reproduce itself in a process known as _____. This is important during cell division as plants _____. Chloroplasts contain the green pigment chlorophyll, an important _____ responsible for bringing about photosynthesis.

[4 MARKS]

8 Plants are made from cells. Draw lines which match the specialised cell types on the left to their main function on the right.

root hair cells		form vessels which conduct simple sugars in sap
guard cells		carry out most photosynthesis in leaves
phloem cells		take up water and dissolved minerals from soil
xylem cells		form vessels which conduct water and dissolved minerals
palisade cells		allow various gases to move into and out of leaves

[5 MARKS]

9 Plants display certain characteristics which demonstrate that they are alive and carry out certain life processes in order to stay alive. These include:

(a) movement, growth and reproduction

(b) respiration, sensitivity, excretion and nutrition

(c) all of the above

[1 MARK]

10 The reproductive cycle of a flowering plant proceeds in five well-defined stages. Draw lines which match the reproductive stages on the left to the description which fits best on the right.

pollination		development of embryo plant
fertilisation		the transfer of pollen from anther to stigma
seed formation		scattering mechanism which helps avoid competition
seed dispersal		appearance of new root and shoot systems
germination		nuclei of male and female sex cells meet and fuse

[5 MARKS]

11 The common causes of ill health in plants include:

(a) mineral deficiencies

(b) invertebrate organisms and plant pathogens

(c) all of the above

[1 MARK]

12 Green plants and other photosynthetic organisms are capable of making their own simple food substances like glucose by photosynthesis. Photosynthesis involves a reaction between carbon dioxide and water. Where does the energy come from to drive this reaction?

[1 MARK]

Structure and function of living organisms: humans and other animals

1 In the diagram of the skeleton shown, label the bones **A** to **R**.

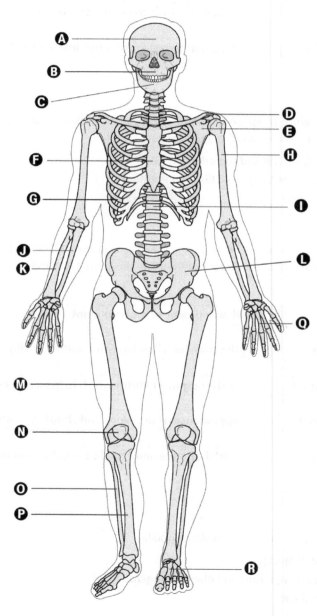

patella

cranium

mandible

scapula

sacrum, ilium and coccyx (bones of the pelvis)

maxilla

radius

femur

carpals, metacarpals and phalanges (bones of the hand)

fibula

sternum

clavicle

humerus

vertebra

rib

ulna

tibia

tarsals, metatarsals and phalanges (bones of the foot)

[18 MARKS]

Complete the following sentences by inserting the most appropriate words.

The human skeleton _____ vital organs. It provides a framework which supports the _____ of individuals and, with the help of muscles, allows humans to _____ upright. The human skeleton also provides attachment for _____ and _____ allowing free movements to take place across joints.

[5 MARKS]

2 How many bones would you expect to find in the skeleton of an average human adult?

(a) 103

(b) 206

(c) 602

[1 MARK]

3 Bone is a living tissue. True or false?

[1 MARK]

4 Refer to the diagram of the human arm.

(a) **Name the muscles labelled A and B.**

(b) **What would you expect to happen as muscle A contracts?**

(c) **What would you expect to happen as muscle B contracts?**

(d) **What are pairs of muscles like A and B called?**

(e) **Where else in the human body might you find a pair of muscles like A and B?**

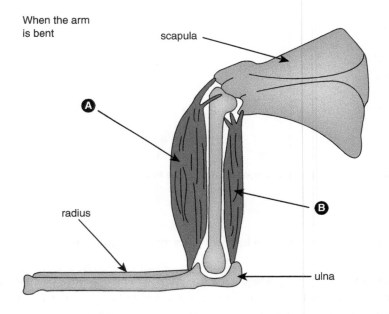

[5 MARKS]

Complete the following sentences by inserting the most appropriate words.

Muscles have the ability to _____ when stimulated by nerves. Almost all movements within the human body, voluntary and involuntary, are caused by muscles, and muscles allow humans to _____ or to get around from one place to another. Muscles are grouped on the basis of structure and function into three types: _____ muscle, _____ muscle and _____ muscle.

[5 MARKS]

5 In the diagram of the human circulatory system shown, label the parts **A** to **J**.

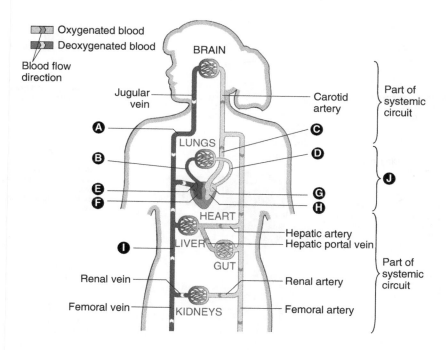

aorta

right ventricle

left ventricle

pulmonary artery

left atrium

right atrium

pulmonary vein

superior vena cava

inferior vena cava

pulmonary circuit

[10 MARKS]

Complete the following sentences by inserting the most appropriate words.

Blood is circulated around the body by the _____, a four-chambered organ that works like two pumps side by side. The human circulatory system transports _____, food substances and _____ all around the body. It gets white blood cells and platelets to where they are needed for the fight against _____ and for _____ wounds.

[5 MARKS]

6 In the diagram of the human digestive system shown, label the parts **A** to **M**.

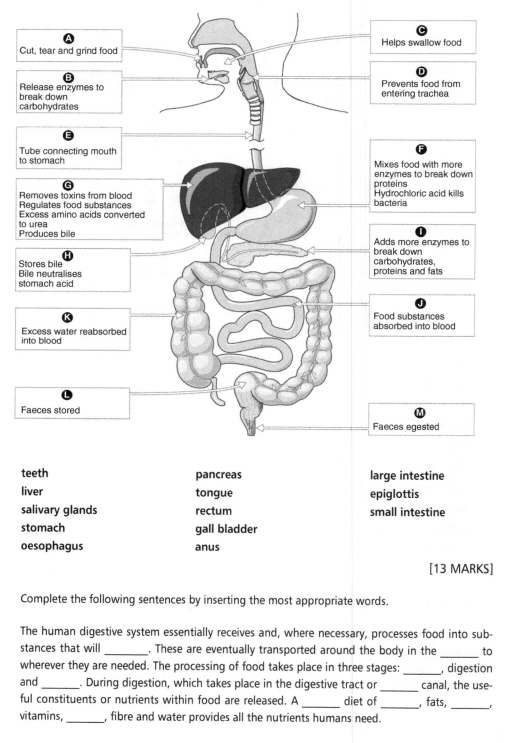

Ⓐ
Cut, tear and grind food

Ⓑ
Release enzymes to break down carbohydrates

Ⓔ
Tube connecting mouth to stomach

Ⓖ
Removes toxins from blood
Regulates food substances
Excess amino acids converted to urea
Produces bile

Ⓗ
Stores bile
Bile neutralises stomach acid

Ⓚ
Excess water reabsorbed into blood

Ⓛ
Faeces stored

Ⓒ
Helps swallow food

Ⓓ
Prevents food from entering trachea

Ⓕ
Mixes food with more enzymes to break down proteins
Hydrochloric acid kills bacteria

Ⓘ
Adds more enzymes to break down carbohydrates, proteins and fats

Ⓙ
Food substances absorbed into blood

Ⓜ
Faeces egested

teeth	pancreas	large intestine
liver	tongue	epiglottis
salivary glands	rectum	small intestine
stomach	gall bladder	
oesophagus	anus	

[13 MARKS]

Complete the following sentences by inserting the most appropriate words.

The human digestive system essentially receives and, where necessary, processes food into sub-stances that will _____. These are eventually transported around the body in the _____ to wherever they are needed. The processing of food takes place in three stages: _____, digestion and _____. During digestion, which takes place in the digestive tract or _____ canal, the use-ful constituents or nutrients within food are released. A _____ diet of _____, fats, _____, vitamins, _____, fibre and water provides all the nutrients humans need.

[9 MARKS]

7 Draw lines which match the teeth on the left to their function on the right:

incisors		tearing food

canines		crushing and grinding hard food

premolars		cutting food

molars		crushing and grinding soft food

[4 MARKS]

8 Label the diagram of the tooth below.

enamel

dentine

pulp

root

crown

[5 MARKS]

Complete the following sentences by inserting the most appropriate words.

The hardest substance in your body is _____, but this can be eroded by _____ which live on your teeth. They feed on _____ and produce _____, which dissolves your teeth causing tooth _____. _____ in toothpaste helps strengthen the _____ to resist acid attack.

[7 MARKS]

9 Complete the following sentences by inserting the most appropriate words.

Humans reproduce sexually. The_____fertilisation of a female _____or sex cell (an egg or ovum) by a male _____ or sex cell (a sperm), usually following sexual intercourse, results in the fusion of cell nuclei and the formation of a_____. This contains all of the genetic information or_____ needed in order to produce a fully formed adult. Eventually, and in the uterus, a human _____ grows within its own environment in a fluid-filled sac or amnion, which protects it. Here, it gets its oxygen and other useful substances from its mother via the _____ and _____ cord. After a gestation period of about _____weeks, female humans give birth to babies who grow and mature in order to complete their own life cycles.

[9 MARKS]

10 In common with all other living organisms, humans display certain characteristics which demonstrate that they are alive, and carry out certain processes in order to stay alive. These are frequently represented in the mnemonic **Mrs Gren.** Which life processes are contained within **Mrs Gren?**

[7 MARKS]

11 In the diagram of the generalised human cell shown, label the parts **A** to **E.**

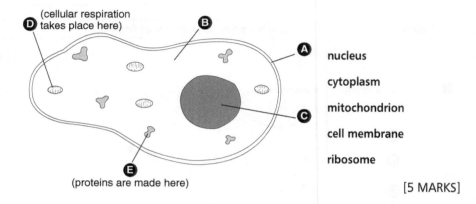

nucleus

cytoplasm

mitochondrion

cell membrane

ribosome

[5 MARKS]

Complete the following sentences by inserting the most appropriate words.

The cell membrane _____ the cytoplasm and nucleus and holds the cell together. Cytoplasm is a jelly-like substance mostly made from _____. The nucleus contains the cell's genetic material or_____. The genetic material has the ability to make identical copies of itself in a process known as _____. Other cell components include mitochondria, where glucose and oxygen react during_____, and ribosomes, where _____ are used to make proteins.

[6 MARKS]

12 With reference to the diagrams of a plant cell and a human cell provided previously (see above and page 22), how are plant and animal cells similar/different?

[2 MARKS]

13 The four main types of pathogen that cause disease in humans are bacteria, viruses, fungi and protoctista. Draw lines which match the pathogens on the left to the types of disease they cause on the right.

bacteria	amoebic dysentery, sleeping sickness, malaria
viruses	sore throats, tuberculosis, typhoid, cholera
fungi	athlete's foot, ringworm, thrush
protoctista	colds, flu, measles, mumps, polio

[4 MARKS]

14 Rearrange the following taxonomic groups in order of size starting with Kingdom.

Kingdom   Species   Order   Genus   Family   Phylum   Class

[1 MARK]

15 Humans are mammals. **True or false?**

[1 MARK]

State the five main characteristics of mammals.

[5 MARKS]

16 Common invertebrate organisms are also animals. Draw lines which match the invertebrate animals on the left to their correct taxonomic group on the right.

woodlice	**arachnids**
worms	**molluscs**
spiders	**crustaceans**
snails	**insects**
butterflies	**annelids**

[5 MARKS]

Interactions and interdependencies

1 Look at the diagram of the food web shown and complete the following tasks.

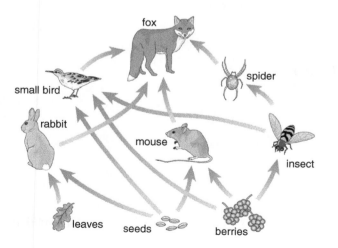

fox
small bird
rabbit
spider
mouse
insect
leaves   seeds   berries

(a) Name one producer.

(b) Name one herbivore and its diet.

(c) Name one omnivore and its diet.

(d) Name one carnivore and its diet.

(e) Write out a complete food chain involving three consumers.

(f) What would you expect to happen to the number of foxes if the number of rabbits increased?

[6 MARKS]

31

2 Draw lines which match the living organism on the left to its position in a simple food chain on the right.

grass		primary consumer (herbivore)
rabbit		producer
fox		secondary consumer (carnivore)

[3 MARKS]

Complete the following sentences using the most appropriate words.

The place where a community of organisms live is called a _____. A community, together with the living and non-living environmental factors which affect how organisms interact and live their lives, is called an _____. All living organisms need food for _____ and the raw materials necessary for healthy growth. _____ plants are _____ and make their own food by photo-synthesis. Animals are _____ and get their food by eating plants and other animals. A food _____ shows how living organisms feed on other living organisms. Complex feeding relation-ships are better shown using a food _____. Micro-organisms which feed on the remains of dead plants and animals and return useful chemicals into the soil are known as _____.

[9 MARKS]

3 Circle the causes of the loss of biodiversity in an ecosystem.

pollution	litter
hole in the ozone layer	increase of human population
traffic	introduction of alien species
intensive farming	global warming

[5 MARKS]

Genetics and evolution

1 Define the following terms.

species

variation

clone

inheritance

mutation

[5 MARKS]

2 Rearrange these structures in order of size starting with the smallest.

cell

gene

nucleus

chromosome

[4 MARKS]

3 Complete the following sentences using the most appropriate words.

In classic genetics, the gene is the basic unit of_____. Genes carry the instructions for different_____. Genes are, in fact, short lengths of_____ and long lengths of _____.

[4 MARKS]

4 How many chromosomes would you expect to find in the typical cells of a healthy human male or female?

(a) 46

(b) 23

(c) 72

[1 MARK]

5 How many chromosomes would you expect to find in the typical sex cells or gametes of a healthy human male or female?

(a) 72

(b) 46

(c) 23

[1 MARK]

6 A human male can roll his tongue (genotype Rr) while his female partner cannot (genotype rr). What are the chances of any children they may have being able to roll their tongues?

(a) 1 in 4

(b) 1 in 2

(c) 3 in 4

[1 MARK]

7 Which of these features could be inherited?

(a) **brown eyes**

(b) **blue hair**

(c) **a scar**

(d) **a pre-disposition to diabetes**

[2 MARKS]

8 Selective breeding involves:

(a) choosing individuals from the same species at random to breed

(b) choosing individuals from different species to breed

(c) choosing individuals from the same species with desirable features to breed

[1 MARK]

9 Draw lines which match the adaptations of the animals to their environment.

polar bear		long eyelashes
camel		large amount of subcutaneous fat
seal		big feet

[3 MARKS]

Chemistry

Materials and their properties including the particulate nature of matter

1 Complete the following sentences by inserting the most appropriate words.

Elements consist of one type of _____ only. There are 92 naturally occurring _____ (there are about 109 in total) and these are grouped according to their similar _____ in the Periodic Table. _____ are formed when two or more substances combine or _____ chemically. _____ are formed when two or more substances are combined physically and the original substances can be _____ relatively easily.

[7 MARKS]

2 In the diagram of the atom shown, label the parts **A** to **C**.

electron

proton

neutron

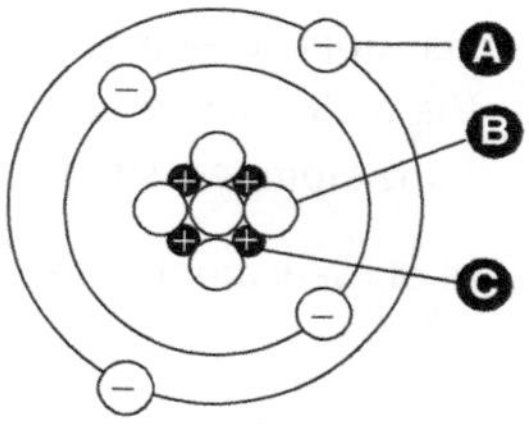

[3 MARKS]

3 Draw lines which match the particle on the left to its best description on the right.

electron		a charge of +1 and a mass of 1 amu
proton		a charge of –1 and a mass of 1/2000 amu
neutron		no charge and a mass of 1 amu

[3 MARKS]

4 Neutrons and protons make up the nucleus of almost all atoms. **True or false?**

[1 MARK]

5 The approximate diameter of one atom is:

(a) 0.001mm

(b) 0.00001mm

(c) 0.0000001mm

[1 MARK]

6 Identify each of the following materials as an element, a compound or a mixture.

hydrogen
ink
sugar
air
copper
honey
pure water
carbon dioxide

[8 MARKS]

7 Atoms combine with each other in a variety of different ways. The type of bond created determines the properties of the substance formed. The main types of bonding are described below.

Ionic: atoms donate or receive electrons forming oppositely charged particles called ions; these particles are then strongly attracted to one another forming giant structures.

Covalent: atoms share electrons with other atoms creating strongly bonded small molecules with weak forces of attraction between the molecules, or atoms share electrons with other atoms creating a giant structure in which all the atoms are strongly bonded together.

Metallic: free electrons from the outer shell of every atom form a 'sea' around the nuclei of the atoms creating a giant structure.

Draw lines which match the substance on the left to its properties and then the properties to the type of bonding that exists on the right.

oxygen	**easily soluble in water** **high melting and boiling point** **solution conducts electricity**	metallic
copper	**high melting and boiling point** **does not conduct electricity** **insoluble in water**	covalent (simple molecules)
diamond	**conducts electricity and heat** **ductile and malleable**	ionic
sodium chloride	**low melting and boiling point** **does not conduct electricity** **does not dissolve in water**	covalent (giant structures)

[8 MARKS]

8 Materials exhibit different physical properties. Some of these are shown in the following table. Complete the table by selecting from the descriptions and examples provided.

Property	Description	Example
Elastic	_____	_____
Plastic	_____	_____
Hard	_____	_____
Tough	_____	_____
Brittle	_____	_____

Descriptions:

breaks easily

deforms when a force is applied but returns to its original shape when force is removed

is permanently deformed as a result of a force acting on it

does not break or tear easily

very difficult to scratch

Examples:

rubber band

play dough

glass

polythene wrapping

diamond

[10 MARKS]

9 Which of the following statements about physical and chemical changes are **true** and which are **false**?

Physical changes:

(a) are usually reversible

(b) are usually irreversible

(c) produce new substances

(d) cause changes in the arrangement of the particles in a substance

Chemical changes:

(a) are usually reversible

(b) are usually irreversible

(c) produce new substances

(d) cause changes in the arrangement of the particles in a substance

[8 MARKS]

10 Draw lines which match the state of matter on the left to its best description on the right.

| solid | | takes the shape of the container it is in, fixed volume, moderate density, very slightly compressible |

| liquid | | no definite shape, no fixed volume, low density, easily compressible |

| gas | | definite shape, fixed volume, moderate to high density, not compressible |

[3 MARKS]

11 In the diagram of the states of matter below, label the changes **A** to **F** which take place between them.

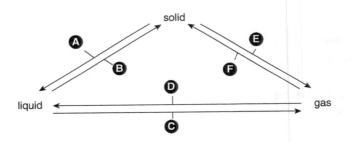

melting

evaporating/boiling

condensing

subliming

solidifying/freezing

reverse subliming

[6 MARKS]

12 Which of the following statements about the change of state of a material from a solid to liquid are **true** and which are **false**?

(a) The particles move faster.

(b) The particles change shape.

(c) The particles lose mass.

(d) The particles move further apart.

(e) The particles gain mass.

(f) The particles have more energy.

(g) The particles are more tightly bonded.

[7 MARKS]

13 When a solid changes to a liquid and then to a gas:

(a) **energy is transferred to the molecules of the substance**

(b) **energy is released**

(c) **no energy is required**

[1 MARK]

14 Identify each of the following as a **chemical** or a **physical** change.

solid carbon dioxide changing into gaseous carbon dioxide

ice changing to water

paper burning

a lump of play dough rolled into a 'wiggly worm'

an egg being boiled

sugar dissolving in a cup of coffee

salt added to an icy path

grass cuttings decomposing in a compost heap

concrete hardening

an iron nail rusting

[10 MARKS]

15 Exothermic reactions occur in chemical reactions when the creation of new bonds requires less energy to form than the bonds that were broken. Endothermic reactions occur in chemical reactions when the creation of new bonds requires more energy than the bonds that were broken. Which of the following examples are exothermic reactions and which are endothermic reactions?

(a) **bicarbonate of soda and vinegar becomes cooler**

(b) **plaster of Paris and water becomes hotter**

(c) **nuclear fission**

(d) **obtaining iron from iron ore**

[4 MARKS]

16 When natural gas burns in oxygen, the resulting compounds or products are carbon dioxide and water (see the word equation below):

$$\text{methane + oxygen} \longrightarrow \text{carbon dioxide + water + heat}$$

The products would have:

(a) **more mass than the reactants**

(b) **less mass than the reactants**

(c) **the same mass as the reactants**

[1 MARK]

17 Draw lines which match the mixture types on the left with the examples on the right.

solid in a solid	air
gas in a liquid	smoke
gas in a gas	Coca Cola
liquid in a gas	clouds
solid in a gas	sand and pebbles
liquid in a liquid	flour in water
solid in a liquid	milk

[7 MARKS]

18 Match the mixtures on the left with the examples of separation technique on the right.

paper clips and sawdust	chromatography
pebbles and sand	dissolving, filtering, evaporating
salt and sand	using a magnet
different coloured inks	sieving
water and alcohol	distillation

[5 MARKS]

19 When a salt dissolves in water the salt crystals:

(a) fill up the spaces between the water molecules

(b) combine with the water to form a new substance

(c) break up and become so small that they are no longer visible

[1 MARK]

20 In each of the following examples, identify the solvent, the solute and the solution.

(a) water, salt, brine

(b) sugar, water, syrup

(c) alcohol, plant oils, perfume

(d) pigment, ink, water

[4 MARKS]

21 You smell air freshener almost as soon as it has been sprayed in a room because:

(a) **the molecules of the air freshener travel faster than the molecules of the air**

(b) **the molecules of the air freshener spread between the molecules of the air**

(c) **the molecules of the air freshener are smaller than the molecules of the air**

[1 MARK]

22 Refer to the graph showing tea in two different cups cooling over time below.

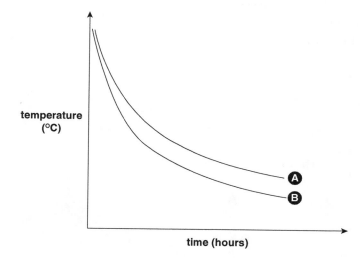

(a) **Which cup keeps the tea hottest for longest, A or B?**

(b) **One cup is made of china and the other polystyrene, which cup is which?**

(c) **Give a reason for your answer.**

(d) **What would happen to the cooling curves if milk were added to the tea?**

(e) **What would happen to the cooling curves if sugar were added to the tea?**

(f) **Explain in energy terms exactly what is happening as the tea cools and the temperature drops. You may wish to consider heat transfers in terms of conduction, convection, radiation and evaporation.**

[6 MARKS]

Earth and atmosphere

1 Complete the table below using the choices provided.

Rock type	Origin	Examples
Igneous	_____	_____
Metamorphic	_____	_____
Sedimentary	_____	_____

Origin:

formed when layers of sediment get buried and crushed under the weight of other layers

formed when heat and pressure completely change existing rocks over long periods of time

formed from the intrusion or extrusion and cooling of molten rock

Examples:

limestone, mudstone, sandstone

granite, basalt

slate, schist, gneiss, marble

[6 MARKS]

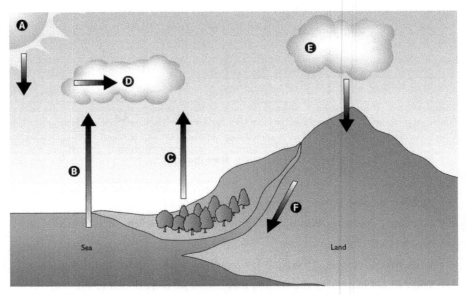

2 Complete the diagram of the water cycle below by labelling the parts **A** to **F**.

evaporation	run-off
precipitation	heat energy from the Sun
transpiration	condensation

[6 MARKS]

3 Coal, oil and natural gas are fuels. Fuels are not forms of energy themselves but potential sources of it. Energy can be released from fuels when they are burned. Burning is an irreversible chemical reaction. Coal, oil and natural gas are examples of non-renewable energy sources. What does the term non-renewable used here mean?

[1 MARK]

4 Energy can also be obtained from the Sun, the wind and waves. These are examples of renewable energy resources. What does the term renewable used here mean?

[1 MARK]

5 Fossils are:

    (a) the petrified remains of animals from a previous geological age

    (b) old plants and animals

    (c) shapes and impression of plants and animals found in rocks

    (d) the petrified remains of different organisms from a previous geological age

[1 MARK]

Physics

Motion and forces

1 Complete the following sentences by inserting the most appropriate words.

A force is a _____, a _____, a twist or a turn. When an object is stationary or moving at a constant speed in a straight line the forces acting on it are said to be _____. Unbalanced forces cause objects to start moving and _____ up, slow down and come to a stop, or change _____. Unbalanced forces also bring about changes in _____. If an object has no _____ force moving it along, it will always slow down and stop because of friction. Friction also includes _____ resistance and _____ resistance. Friction always increases as the speed of a moving object increases. Friction can also be useful. Without friction, standing up, riding a bicycle or driving around in cars would be very _____.

[10 MARKS]

2 Identify and draw, using force arrows, the balanced forces operating in each of the examples below.

(a) **a book resting on a table**

(b) **a car travelling steadily along a motorway**

(c) **an oil tanker at sea**

(d **a jet aircraft cruising at altitude**

[8 MARKS]

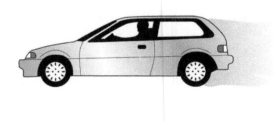

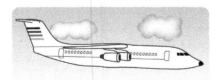

3 Examples of contact forces include:

(a) **pushes and pulls**

(b) **twists and turns**

(c) **all of the above**

[1 MARK]

4 Examples of non-contact forces include:

(a) **gravity and magnetism**

(b) **electrostatic attraction and repulsion**

(c) **all of the above**

[1 MARK]

5 In what units are forces measured?

(a) **kilograms (kg)**

(b) **newtons (N)**

(c) **pascals (Pa)**

[1 MARK]

6 Describe what you would expect to see happen when a feather and a hammer are dropped from the same height at the same time by a person standing on the surface of the Earth. Describe what you would expect to see happen if the same experiment took place on the surface of the Moon. (This experiment actually took place during one of several Apollo missions to the Moon.)

[2 MARKS]

7 Complete the following sentences by inserting the most appropriate words.

Mass and weight are separate things. The mass of an object is simply the amount of _____ in it. Mass is measured in _____ (kg). The mass of an object remains the same regardless of whether it is found on Earth or in space. Interestingly, any two objects with mass exert a _____ on each other but this is only noticeable when one of the objects is particularly massive. This force of attraction between all masses is called gravity. Weight is a force. As a result, weight is measured in _____ (N). The weight of an object changes depending on where it is in the Universe. All objects on the surface of the Earth are pulled towards it with a force of about 10 N/kg. The Moon is much less massive than the Earth. All objects on the surface of the Moon are pulled towards it with a force of about 1.6 N/kg. An object on the surface of the Earth therefore weighs _____ than the same object on the surface of the Moon even though it has exactly the _____ mass.

[6 MARKS]

8 The force of gravity on the surface of the Earth is about 10 N/kg. How much would a person of mass 90 kg weigh?

[1 MARK]

45

9 The force of gravity on the surface of the Moon is about 1.6 N/kg. How much would a person of mass 90 kg weigh?

[1 MARK]

10 Gravity forces can act between two objects with mass at a distance. **True or false?** Give an example to justify your answer.

[2 MARKS]

11 Complete the following sentences by inserting the most appropriate words.

The movement of an object can be described in terms of its speed or how _____ it is going. The speed of an object can be calculated easily if we know the _____ it travels and the _____ taken to travel that distance. Speed is usually measured in _____ (m/s). The movement of an object can also be described in terms of its velocity. The term velocity should be used in preference to speed when the _____ in which an object is moving is given. Objects do not always travel at a constant speed or velocity, however. They can always speed up or change direction. Objects which speed up, change direction or do both at the same time are said to _____.

[6 MARKS]

12 In the middle of a race, a cyclist travels along one 540 m stretch of straight road in 45 seconds. Calculate the speed of the cyclist. At that speed, how long would it take the cyclist to travel a further 120 m? Later in the same race, the cyclist registers a constant speed of 15 m/s for 2 minutes. How far does the cyclist travel in that time?

[3 MARKS]

Sound

1 In the diagram of the human ear shown, label the parts **A** to **M**.

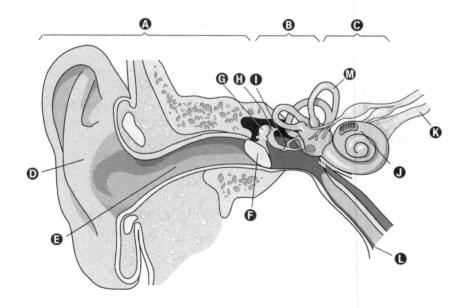

ear drum	Eustachian tube
outer ear	inner ear
ear canal	pinna
cochlea	auditory nerve
anvil	middle ear
stirrup	semi-circular canals
hammer	

Complete the following sentences by inserting the most appropriate words.

Ears allow us to hear. We have two ears in order to help locate sound sources accurately. Sounds entering the ears cause the ear drums to _____, which in turn force the three small bones of the middle ear to move. Specialised cells within the cochlea (sensitive to vibration and movement) change sound energy to electrical energy. Electrical impulses travel via the auditory nerve to the brain where they are _____ and interpreted as sound. Unwanted or disagreeable sound is called _____. Exposure to particularly loud sounds even for short periods of time can damage the ear drum and lead to partial or complete _____. The ears are never 'switched off'. The Eustachian tube in each ear is responsible for maintaining _____ balance between the middle ear and the outside world. The semi-circular canals are concerned with _____ and orientation and not hearing.

[6 MARKS]

2 Sound travels in air at:

(a) **about 330 m/s**

(b) **about 330 km/s**

(c) **about 330 cm/s**

[1 MARK]

3 Underline the most appropriate word found in brackets.

Sounds generally travel (faster/more slowly) in solids and liquids than in air.

[1 MARK]

4 Underline the most appropriate word found in brackets.

Sound travels in waves. These waves are described as (transverse/longitudinal) waves.

[1 MARK]

5 Complete the following sentences by inserting the most appropriate words.

Sound is a form of _____. Sounds travel outwards from a _____ source in waves. Sounds will travel in solids, liquids and gases but not in a _____. Some sounds are louder or quieter than others and differ in _____. Some sounds are higher or lower in pitch than others and differ in _____. Noise levels are often measured in _____ (dB). Sound waves can also be reflected. Sound reflections are referred to as _____.

[7 MARKS]

6 Consider the following sources of sound. For each, identify the source of vibration. In the case of the guitar, identify three ways in which the pitch of the sound produced can be changed and one way in which the volume of the sound can be changed.

(a) piano

(b) drum

(c) recorder

(d) guitar

[8 MARKS]

Light

1 In the diagram of the human eye shown, label the parts **A** to **K**.

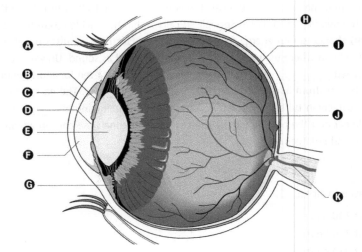

eyelid	pupil	aqueous humour
lens	optic nerve	vitreous humour
retina	cornea	sclera (white of eye)
iris	ciliary muscle	

[11 MARKS]

Complete the following sentences by inserting the most appropriate words.

Eyes allow us to see. Each eye has a series of _____ which allow movement within the retaining eye socket. Two eyes provide _____ vision, which provides depth perception and distance judgement. Light enters the eye through the cornea and lens which focus the light rays onto a _____-sensitive layer called the retina. Specialised cells within the retina called _____ (sensitive to colour) and _____ (sensitive to 'greys') change light energy to electrical energy. Electrical impulses travel via the optic nerve to the brain where they are _____ and interpreted as sight. Between the cornea and the lens is the iris. The iris determines the _____ of the eye. At the centre of the iris is the pupil. The pupil changes size in response to the amount of _____ entering the eye. This means that in areas with _____ light the pupil will be _____ than in darker areas.

[10 MARKS]

2 Underline the most appropriate words in brackets.

Short-sightedness or myopia results in light rays being focused 'short' of the retina and within the eyeball itself. This can be corrected using a (converging/diverging) or concave lens. Long-sightedness or hyperopia results in light rays being focused beyond the retina and 'behind' the eyeball itself. This can be corrected using a (converging/diverging) or convex lens.

[2 MARKS]

3 Which of the following sources of light are primary and which are secondary?

(a) a torch

(b) the Moon

(c) a burning candle

(d) the Sun

[4 MARKS]

4 Light travels in a vacuum at:

(a) 300 000 km/s

(b) 300 000 m/s

(c) 300 000 mph

[1 MARK]

5 Underline the most appropriate word found in brackets.

Light travels in waves. These waves are described as (transverse/longitudinal) waves.

[1 MARK]

6 Complete the following sentences by inserting the most appropriate words.

Light is a form of _____. Light travels in straight lines from a source unless prevented from doing so. Light is a small part of what is referred to as an _____ spectrum of waves, which include gamma rays, X-rays, UV, IR, microwaves and radio waves. Light waves have some important features that can be measured: the _____, which determines the colour of the light, the _____, or the number of waves that pass every second, and the _____, which determines the intensity or brightness of the light.

[5 MARKS]

7 The primary colours of light are:

(a) red, blue and yellow

(b) red, blue and green

(c) red, green and yellow

[1 MARK]

8 The primary colours of artists' pigments are:

(a) **red, green and yellow**

(b) **red, blue and green**

(c) **red, blue and yellow**

[1 MARK]

9 A beam of 'white' light passing through a glass prism is 'bent' and split into its component colours. This effect is known as dispersion. List the seven colours you might expect to see in the 'spectrum' produced.

[1 MARK]

10 The colour of an object is actually the colour or the wavelength of the light it reflects. All other colours or colour wavelengths are absorbed.

(a) **What colour wavelengths do green objects absorb?**

(b) **Why do some objects appear white and others appear black?**

[3 MARKS]

11 Underline the most appropriate words found in brackets.

Shadows are formed when light is blocked. When the light from a projector is blocked by an object, the shadow formed on a wall, for example, can be made (bigger/smaller) by increasing the distance between the object and the screen or by decreasing the distance between the projector and the object. Some shadows are 'black'. Some shadows appear with a dark central area and a fuzzy, grey outline. The dark part of the shadow is known as the (umbra/penumbra). The fuzzy, grey outline is known as the (umbra/penumbra).

[3 MARKS]

12 Look at the diagram below. Complete the ray diagram to indicate how and where the image of the candle appears to the observer.

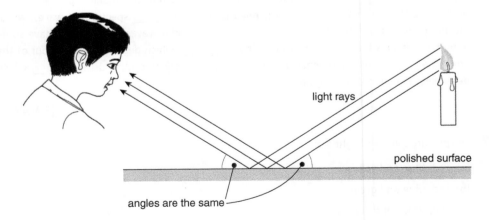

light rays

polished surface

angles are the same

[2 MARKS]

13 Why are polished surfaces better at reflecting light than rough surfaces?

[1 MARK]

14 Match the words on the left with their description on the right.

transparent		blocks light and images completely

translucent		allows light to pass and objects to be seen clearly

opaque		allows light to pass but objects appear blurred

[3 MARKS]

Electricity and magnetism

1 Look at the picture of a simple circuit shown here and complete the following tasks. The filament lamps or bulbs are identical.

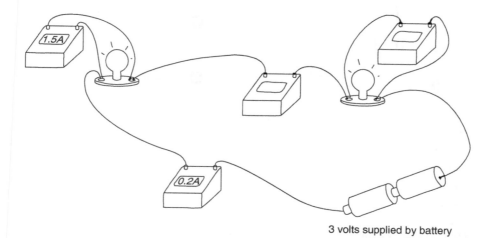

3 volts supplied by battery

(a) This is a series circuit. True or false?

(b) Underline the most appropriate words in brackets. Unscrewing one bulb from its holder will cause the other bulb to (glow brighter/glow dimmer/glow about the same/go out).

(c) The current flowing in the circuit as shown by one ammeter is 0.2A. What reading would you expect the other ammeter to show?

(d) The voltage (or potential difference) across one of the bulbs as shown by one voltmeter is 1.5V. What reading would you expect the other voltmeter to show?

(e) Use Ohm's Law (V = IR) to calculate the resistance, R, of each of the two bulbs.

(f) Draw a circuit diagram of the picture using standard symbols.

(g) Use the circuit diagram to describe what is happening in the circuit in terms of energy and energy transfer.

(h) Imagine adding a third identical bulb to the circuit.

- The three identical bulbs will be equally bright. True or false?
- The three identical bulbs will be as bright as the two bulbs in the previous series circuit. True or false?

Briefly explain why, using the concept of 'resistance'.

(i) Draw a circuit diagram which shows the two original bulbs connected in parallel.

(j) Refer to the parallel circuit.

- How would you expect the bulbs to appear in terms of their brightness compared with the series circuit?
- What will happen if you remove one of the bulbs?

[13 MARKS]

2 In the diagram of the circuit symbols shown, identify the components **A** to **G**.

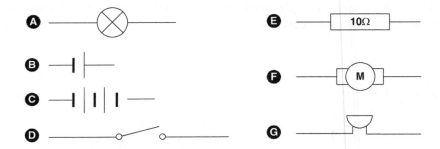

motor

resistor

battery

switch

buzzer

cell

filament lamp or bulb

[7 MARKS]

3 Define the following terms.

conductor

insulator

current

voltage

resistance

power

[6 MARKS]

4 Under normal circumstances, which of the following materials are good electrical conductors, poor electrical conductors or insulators?

(a) copper

(b) water

(c) rubber

(d) aluminium

(e) gold

(f) wood

(g) skin

(h) air

[8 MARKS]

5 Complete the following sentences by inserting the most appropriate words.

Some materials are magnetic. That means that they are both _____ to magnets and can be _____. Magnetic materials include the metals _____ and _____ as well as nickel and cobalt. All magnets have _____ and _____ poles. The rule of magnets states that _____ poles attract while _____ poles _____. That means that when _____ poles of magnets are brought close together it is possible to feel an _____ force between them. When _____ poles of magnets are brought close together it is possible to feel a _____ force between them. A simple compass is nothing more than a freely moving magnet which aligns itself with the Earth's magnetic field. The end of the magnet which points north is referred to as the _____-seeking pole. The rule of magnets tells us, therefore, that what we refer to as geographical north is actually the Earth's magnetic _____ pole.

[15 MARKS]

6 A simple electromagnet can be made by winding a length of wire around an iron nail. When connected to a battery, a current flows through the wire creating a magnetic field around it. The magnetic field strength is increased by the presence of the nail. List up to three changes that could be made to alter the magnetic field strength of this simple electromagnet.

[3 MARKS]

Space physics

1 Complete the following sentences by inserting the most appropriate words.

The Universe is, quite literally, everything that exists: _____ (from atoms and molecules to stars and galaxies), _____ (visible light together with the rest of the electromagnetic spectrum) and _____ (the vast emptiness within and between galaxies). The Universe is about 12 _____ years old and most probably emerged from an explosive event referred to as the _____. The Universe has been growing in size or _____ ever since.

[6 MARKS]

2 Complete the following sentences by inserting the most appropriate words.

Galaxies are assemblages of _____, nebulae and other interstellar materials. A typical galaxy contains more than _____ billion stars and measures about _____ light years across. Galaxies are classified into four main groups depending on their appearance: _____, barred spirals, _____ and irregulars. Galaxies are not randomly scattered throughout the Universe; they occur in clusters: _____ clusters of hundreds or thousands of galaxies and _____ clusters of a few tens. Our own Sun is located within the Orion Arm of what is referred to as the _____ galaxy, one of about 30 other galaxies known as the Local Group.

[8 MARKS]

3 Rearrange the following in order of decreasing size starting with the Universe.

Universe	Solar System	local group of galaxies
Earth–Sun–Moon system	Earth	Milky Way
	Moon	Sun

[1 MARK]

4 Draw lines which match the contents of the Solar System on the left to their best description on the right.

Sun	chunks of ice and other material often seen with a tail
planets	natural satellites which orbit planets
moons	a star (ball of hot, glowing gas)
asteroids	lumps of rock often referred to as minor planets
comets	small particles of dust and rock fragments
meteoroids	rocky and gassy objects which orbit the Sun

[6 MARKS]

5 Refer to the Solar System.

(a) List the nine known planets which orbit the Sun in order starting with Mercury.

(b) Which planets are known as the terrestrial or rocky planets?

(c) Which planets are known as the Jovian planets or gas giants?

(d) Mercury, being nearest to the Sun, is the hottest planet. True or false?

(e) Which is the largest planet?

(f) Which planet has the most moons?

(g) Which planets have no moons at all?

(h) Which planets have rings?

(i) The Solar System formed about 4.6 billion years ago. True or false?

[9 MARKS]

6 Correctly label features **A** to **F** of the day and night cycle in the diagram shown.

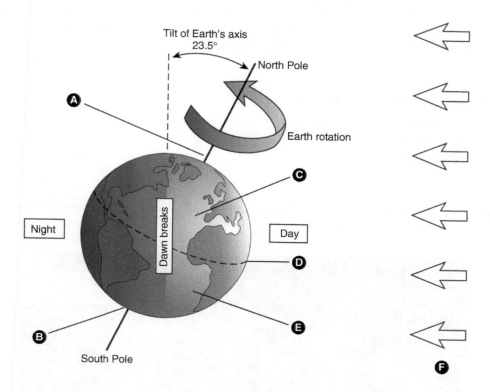

light from the Sun

where more than 12 hours of daylight are experienced

where the Sun never 'rises'

where exactly 12 hours of daylight are experienced

where the Sun never 'sets'

where less than 12 hours of daylight are experienced

[6 MARKS]

7 The day and night cycle is caused by the rotation of the Earth about its axis. The time from one 'sunrise' to the next is:

(a) approximately 24 hours

(b) 365.25 days

(c) about one month

[1 MARK]

8 With reference to the seasons, complete the following sentences by inserting the most appropriate words and by circling the most appropriate words found in brackets.

The _____ of the Earth's axis relative to the plane of its orbit around the Sun causes the seasons. In the UK, the year-long cycle of seasons includes spring, summer, autumn and winter. In June, the _____ hemisphere is tilted _____ the Sun and experiences summer while the

_____ hemisphere is tilted away and experiences _____. The effects are dramatic. In the UK, for example, summer days are (long/short), the Sun 'rises' (high/low) above the horizon so the Sun's rays reach the surface of the Earth at a (high/low) angle, and the Earth is (heated/cooled) by the Sun for (more/less) than 12 hours. The Sun's heating effect is (more/less) efficient and summers are warm. In December the opposite occurs.

[11 MARKS]

9 The Earth orbits the Sun once every:

(a) **four years**

(b) **24 hours**

(c) **365.25 days**

[1 MARK]

10 Correctly draw the phases of the Moon as they would be seen from the Earth in the boxes labelled **A** to **H** in the following diagram.

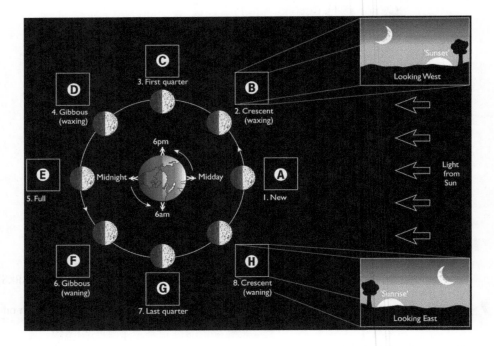

[8 MARKS]

11 The Moon orbits the Earth about:

(a) **once every 24 hours**

(b) **once a month**

(c) **once a year**

[1 MARK]

Making sense of your test results

How well did you do? Determine a separate percentage score for Biology, Chemistry and Physics. Determine an overall score for the test. Remember that your percentage score is relative to the nature of the material tested and the time at which the test took place.

		Score
Biology	Marks [max 249]	_____ %
Chemistry	Marks [max 120]	_____ %
Physics	Marks [max 238]	_____ %
Overall	Marks [max 607]	_____ %

Consider the following divisions against which your separate and overall test scores can be measured. The scale is based upon our own experiences of testing trainees in this way over the past few years. It should be used for guidance and relative improvement only and not taken as some sort of absolute test measure.

80–100% In the areas tested, your score is very good and indicates that you probably exceed the level expected of a non-science specialist. Well done.

60–80% In the areas tested, your score is good and indicates that you probably meet the level expected of a non-science specialist. Some attention is necessary locally, certainly in the weaker questions. With these marks you probably have little to worry about.

50–60% In the areas tested, your score is adequate and probably indicates that you are moving towards the level expected of a non-science specialist. However, attention is necessary throughout. Just like perceived competence, you are probably in good company and with a little effort you will be up there with the best of them.

0–50% In the areas tested, your score is probably a bit on the low side. But what did you expect? Use the test positively to target the bits you need to work on and really work on them. Remember, you only have to get there by the end of your training. Your score would not concern us at this stage so don't let it concern you.

A useful tip would be to take a break from testing for now. Use the test questions as a guide for some revision. Come back to the test again later and see how much progress you have made.

Part 5: Answers to test questions including children's common misconceptions

Biology
Structure and function of living organisms: plants

1 **A** flower; **B** leaf; **C** stem; **D** tap root; **E** lateral root; **F** root hair; **G** shoot system; **H** root system

Roots **anchor** plants firmly in the ground. They are also responsible for the uptake of **water** and **minerals** from the soil. Stems hold plants upright, spread out leaves for **photosynthesis** and elevate flowers for **pollination**. Most leaves are green due to the presence of **chlorophyll**. Flowers are the structures of most plants that are responsible for **reproduction.**

2 **A** petal; **B** stigma; **C** style; **D** ovary with ovules; **E** anther; **F** filament; **G** sepal; **H** receptacle

Flowers usually consist of five main elements (though these are not always present or immediately obvious): often brightly coloured and scented petals which attract **pollinators**, papery sepals which **protect** the flower in bud, male and female reproductive organs which work together to ensure the **continuity** of each species, and a receptacle which **supports** the weight of everything.

3 (c) the anther and filament

4 (a) the stigma, style and ovary

5 **A** waxy cuticle; **B** upper epidermal cells; **C** palisade mesophyll cells; **D** spongy mesophyll cells; **E** lower epidermal cells; **F** guard cell; **G** stoma; **H** 'vein' (phloem and xylem vessels)

The leaves of many plants are large and **flat** in order to trap sunlight and to make the process of photosynthesis particularly efficient. A cross-section through a leaf blade or **lamina** reveals several layers including one consisting of palisade cells within the palisade mesophyll where most photosynthesis takes place. The rate of photosynthesis is affected by several factors including **light** intensity, air **temperature**, the concentration of **carbon dioxide** in the atmosphere, and **water** availability.

6 $$\text{carbon dioxide + water} \xrightarrow[\text{(chlorophyll)}]{\text{(energy in sunlight)}} \text{glucose + oxygen}$$

(*Note:* 1 mark for each side of the equation and 1 mark for sunlight and chlorophyll)

7 **A** cellulose cell wall; **B** cell membrane; **C** cytoplasm; **D** nucleus; **E** sap-filled vacuole; **F** chloroplast with chlorophyll; **G** ribosome; **H** mitochondrion

The nucleus contains the plant cell's genetic material or **DNA**, which ultimately determines what type of cell it is and controls what it does. The genetic material also has the ability to reproduce itself in a process known as **replication**. This is important during cell division as plants **grow**. Chloroplasts contain the green pigment chlorophyll, an important **enzyme** responsible for bringing about photosynthesis.

8 root hair cells   to   take up water and dissolved minerals from soil

    guard cells   to   allow various gases to move into and out of leaves

    phloem cells   to   form vessels which conduct simple sugars in sap

    xylem cells   to   form vessels which conduct water and dissolved minerals

    palisade cells   to   carry out photosynthesis in leaves

9 (c) all of the above

10 pollination   to   the transfer of pollen from anther to stigma

    fertilisation   to   nuclei of male and female sex cells meet and fuse

    seed formation   to   development of embryo plant

    seed dispersal   to   scattering mechanism which helps avoid competition

    germination   to   appearance of new root and shoot systems

11 (c) all of the above

12 The energy which drives photosynthesis comes from the Sun.

Common misconceptions including simple explanations in brackets
Children may think that:

- plants are not living things (plants are alive even though some of the characteristics of living things, e.g. movement and sensitivity, are less obvious in plants than in animals);

- living things are those that move (children often over-emphasise movement as a criterion for classification of objects as 'living' and do not take into account all seven conditions required – Mrs Gren);

- trees, grass, vegetables and weeds are not plants (children do not classify systematically and their criteria for classification lack discrimination; there are many types of plants and not all plants have the same structures);

- seeds require light and soil as well as warmth and water to germinate (most seeds only require oxygen, warmth and water to germinate; seedlings, once germinated, require light, as well as minerals from soil);

- plants take all the things they need from the soil through their roots in order to grow (water and mineral salts are taken in through the roots from the soil; air is taken in through the leaves, and the chloroplasts in the leaves absorb the Sun's energy and use the carbon dioxide in the air together with water to produce glucose in the process of photosynthesis);

- flowers are for human enjoyment and to look pretty (children may have anthropocentric ideas about plants; flowers are the sexual reproductive structure of a flowering plan; most non-flowering plants reproduce sexually via spores).

Useful references
Allebone, B. (1995) Children's ideas about plants. *Primary Science Review*, 39: 20–23.

Barker, M. (1995) A plant is an animal standing on its head. *Journal of Biological Education*, 29(30): 201–208.

Bianchi, L. (2000) So what do you think a plant is? *Primary Science Review*, 61: 15–17.

Huxham, M., Welsh, A., Berry, A. and Templeton, S. (2006) Factors influencing primary school children's knowledge of wildlife. *Journal of Biological Education*, 41(1): 9–12.

Jewell, N. (2002) Examining children's models of seed. *Journal of Biological Education*, 36(3), 116–121.

Osborne, J., Wadsworth, P. and Black, P. (1992) *Processes of life: Primary SPACE Research Report*. Liverpool: LUP.

Russell, T. and Watt, D. (1990) *Growth: Primary SPACE Research Report*. Liverpool: LUP.

Schussler, E. (2008) From flowers to fruits: how children's books represent plant reproduction. *International Journal of Science Education*, 30(12): 1677–1696.

Tunnicliffe, S.D. and Reiss, M.J. (2000) Building a model of the environment: how do children see plants? *Journal of Biological Education*, 34(4): 172–177.

Structure and function of living organisms: humans and other animals

1 **A** cranium; **B** maxilla; **C** mandible; **D** clavicle; **E** scapula; **F** sternum; **G** rib; **H** humerus; **I** vertebra; **J** ulna; **K** radius; **L** sacrum, ilium and coccyx (bones of the pelvis); **M** femur; **N** patella; **O** fibula; **P** tibia; **Q** carpals, metacarpals and phalanges (bones of the hand); **R** tarsals, metatarsals and phalanges (bones of the foot)

The human skeleton **protects** vital organs. It provides a framework which supports the **weight** of individuals and, with the help of muscles, allows humans to **stand** upright. The human skeleton also provides attachment for **muscles** and **tendons** allowing free movements to take place across joints.

2 (b) 206

3 True

4 (a) **A** biceps; **B** triceps

(b) the arm bends or flexes at the elbow

(c) the arm straightens or extends at the elbow

(d) antagonistic pairs – they work together so when one relaxes the other contracts

(e) the leg

Muscles have the ability to **contract** when stimulated by nerves. Almost all movements within the human body, voluntary and involuntary, are caused by muscles, and muscles allow humans to **locomote** or to get around from one place to another. Muscles are grouped on the basis of structure and function into three types: **skeletal** muscle, **smooth** muscle and **cardiac** muscle.

5 **A** superior vena cava; **B** pulmonary artery; **C** aorta; **D** pulmonary vein; **E** right atrium; **F** right ventricle; **G** left atrium; **H** left ventricle; **I** inferior vena cava; **J** pulmonary circuit

Blood is circulated around the body by the **heart**, a four-chambered organ that works like two pumps side by side. The human circulatory system transports **blood**, food substances and **heat** all around the body. It gets white blood cells and platelets to where they are needed for the fight against **disease** and for **healing** wounds.

6 **A** teeth; **B** salivary glands; **C** tongue; **D** epiglottis; **E** oesophagus; **F** stomach; **G** liver; **H** gall bladder; **I** pancreas; **J** small intestine; **K** large intestine; **L** rectum; **M** anus

The human digestive system essentially receives and, where necessary, processes food into substances that will **dissolve**. These are eventually transported around the body in the **blood** to wherever they are needed. The processing of food takes place in three stages: **ingestion**, digestion and **egestion**. During digestion, which takes place in the digestive tract or **alimentary** canal, the useful constituents or nutrients within food are released. A **balanced** diet of **carbohydrates**, fats, **proteins**, vitamins, **minerals**, fibre and water provides all the nutrients humans need.

7

incisors	to	cutting food
canines	to	tearing food
premolars	to	crushing and grinding soft food
molars	to	crushing and grinding hard food

8 **A** enamel; **B** dentine; **C** pulp; **D** root; **E** crown

The hardest substance in your body is **enamel** but this can be eroded by **bacteria**, which live on your teeth. They feed on **sugars** and produce **acid**, which dissolves your teeth causing tooth **decay**. **Fluoride** in toothpaste helps strengthen the **enamel** to resist acid attack.

9 Humans reproduce sexually. The **internal** fertilisation of a female **gamete** or sex cell (an egg or ovum) by a male **gamete** or sex cell (a sperm), usually following sexual intercourse, results in the fusion of cell nuclei and the formation of a **zygote** containing all of the genetic information or **DNA** needed in order to produce a fully formed adult. Eventually, and in the uterus, a human **foetus** grows within its own environment in a fluid-filled sac or amnion, which protects it. Here, it gets its oxygen and other useful substances from its mother via the **placenta** and **umbilical** cord. After a gestation period of about **40** weeks, female humans give birth to babies who grow and mature in order to complete their own life cycles.

10 **M** movement; **R** reproduction; **S** sensitivity; **G** growth; **R** respiration; **E** excretion; **N** nutrition

11 **A** cell membrane; **B** cytoplasm; **C** nucleus; **D** mitochondrion; **E** ribosome

The cell membrane **encloses** the cytoplasm and nucleus and holds the cell together. Cytoplasm is a jellylike substance mostly made from **water**. The nucleus contains the cell's genetic material or **DNA**. The genetic material has the ability to make identical copies of itself in a process known as **replication**. Other cell components include mitochondria, where glucose and oxygen react during **respiration**, and ribosomes, where **amino acids** are used to make proteins.

12 Generally, plant and animal cells both possess a cell membrane, cytoplasm, a nucleus and other organelles including mitochondria and ribosomes. Unlike most plant cells, animal cells do not possess a cellulose cell wall, a sap-filled vacuole or, as in the case of green tissue plant cells, chloroplasts containing chlorophyll.

13

bacteria	to	sore throats, tuberculosis, typhoid, cholera
viruses	to	colds, flu, measles, mumps, polio
fungi	to	athlete's foot, ringworm, thrush
protoctista	to	amoebic dysentery, sleeping sickness, malaria

14 Kingdom, Phylum, Class, Order, Family, Genus, Species

15 True

1. suckle their young

2. have hairy bodies

3. give birth to living young

4. are vertebrates

5. are warm blooded

16

woodlice	to	crustaceans
worms	to	annelids
spiders	to	arachnids
snails	to	molluscs
butterflies	to	insects

Common misconceptions including simple explanations in brackets
Children may think that:

- anything that moves is alive, e.g. the Moon, clocks, fires, cars, bicycles (animism is the ascription of animal characteristics to items that are not animals);

- animals have happy faces (anthropomorphism is the ascription of human characteristics and feelings to other objects and living things);

- humans are not animals (children often classify human beings as a separate category to other animals);

- food and drink are taken into the human body and travel through it separately, blood is not contained in blood vessels, and the bones of the skeleton are not connected together (children's knowledge of the structure and function of the systems within the human body is generally not well developed).

Useful references
Bell, F.B. (1981) When is an animal, not an animal? *Journal of Biological Education*, 15(3): 213–218.

Bell, B. and Barker, M. (1982) Towards a scientific concept of 'animal'. *Journal of Biological Education*, 16(23): 197–200.

Braund, M. (1991) Children's ideas in classifying animals. *Journal of Biological Education*, 25(2): 103–110.

Braund, M. (1998) Trends in children's concepts of vertebrate and invertebrate. *Journal of Biological Education*, 32(2): 112–118.

Byrne, J. (2011) Models of micro-organisms: children's knowledge and understanding of micro-organisms from 7 to 14 years old. *International Journal of Science Education*, 33(14): 1927–1961.

Byrne, J., Grace, M. and Hanley, P. (2009) Children's anthropomorphic and anthropocentric ideas about micro-organisms: do they affect learning? *Journal of Biological Education Special Issue*, 44(1): 37–43.

Cuthbert, A.J. (2000) Do children have a holistic view of their internal body maps? *School Science Review*, 82: 25–32.

Patrick, P., Byrne, J., Tunnicliffe, S., Asunta, S., Carvalho, S., Havu-Nuutinen S., Sigurjónsdóttir, H., Óskarsdóttir, G. and Tracana, R. (2013) Students (ages 6, 10, and 15 years) in six countries knowledge of animals. *Nordina*, 9(2): 18–32.

Prokop, P., Prokop, M., Tunnicliffe, S.D. and Diran, C. (2007) Children's ideas of animals' internal structures. *Journal of Biological Education*, 41(2): 62–67.

Reiss, M. and Tunnicliffe, S.D. (1999) Children's knowledge of the human skeleton. *Primary Science Review*, 60: 7–10.

Reiss, M.J., Tunnicliffe, S.D., Andersen, A.M., Bartoszeck, A., Carvalho, G., Chen, S., Jarman, R., Jonsson, S., Manokore, V., Marchenko, N., Mulemwa, J., Novikova, T., Otuka, J., Teppa, S. and VanRooy, W. (2002) An international study of young people's drawings of what is inside themselves. *Journal of Biological Education*, 36(2): 58–64.

Shepardson, D.P. (2002) Bugs, butterflies, and spiders: children's understanding about insects. *International Journal of Science Education*, 24(6): 627–643.

Teixeira, F.M. (2000) What happens to the food we eat? Children's conceptions of the structure and function of the digestive system. *International Journal of Science Education*, 22(5): 507–520.

Tunnicliffe, S.D. and Reiss, M.J. (1999) Building a model of the environment: how do children see animals? *Journal of Biological Education*, 33(3): 142–148.

Interactions and interdependencies

1   (a) From the diagram, the leaves, seeds and berries are the 'producers'. Strictly speaking, however, the green plants (e.g. the trees and shrubs) they come from are actually the producers, for it is the green plants that produce their own food by photosynthesis and make the leaves, seeds and berries available.

(b) The rabbit is a herbivore; it eats leaves.

(c) The small bird is an omnivore; it eats insects and berries.

(d) The fox is a carnivore; it eats almost all of the other animals present.

(e) berries ⟶ insect ⟶ small bird ⟶ fox

(f) An increase in the number of rabbits potentially provides the foxes with more food. If this situation were to persist, such favourable conditions would support a larger fox population. Fox numbers would therefore be expected to increase too.

2

grass	to	producer
rabbit	to	primary consumer (herbivore)
fox	to	secondary consumer (carnivore)

3 The place where a community of organisms lives is called a **habitat**. A community, together with the living and non-living environmental factors which affect how organisms interact and live their lives, is called an **ecosystem**. All living organisms need food for **energy** and the raw materials necessary for healthy growth. **Green** plants are **autotrophs (producers)** and make their own food by photosynthesis. Animals are **heterotrophs (consumers)** and get their food by eating plants and other animals. A food **chain** shows how living organisms feed on other living organisms. Complex feeding relationships are better shown using a food **web**. Micro-organisms which feed on the remains of dead plants and animals and return useful chemicals into the soil are known as **decomposers**.

4 pollution; increase of human population; introduction of alien species; intensive farming; global warming

Common misconceptions including simple explanations in brackets
Children may think that:

- food chains only involve predators and prey (the roles of the Sun as the ultimate source of energy and plants as primary producers are frequently ignored);

- organisms that are higher in a food web eat everything that is below them (there are different feeding relationships in a food web; a simple version is that herbivores eat plants, primary predators eat the herbivores and secondary predators eat the primary predators);

- food chains constitute the relationship in an ecosystem (a community of organisms in a habitat form a complex food web that more accurately depicts the energy flow in an ecosystem).

Useful references
Leach, J., Driver, R., Scott, P. and Wood-Robinson, C. (1996) Children's ideas about ecology 2: ideas found in children aged 5–16 about the cycling of matter. *International Journal of Science Education*, 18(1): 19–34.

Leach, J., Driver, R., Scott, P. and Wood-Robinson, C. (1996) Children's ideas about ecology 3: ideas found in children aged 3–16 about the interdependency of organisms. *International Journal of Science Education*, 18(2): 129–141.

Genetics and evolution

1 **Species**: a group of living organisms which share a wide range of common characteristics and can breed together to produce fertile offspring

Inheritance: those characteristics or features which are received as a result of the genetic make-up of biological parents

Variation: how living organisms of the same species look or behave differently from each other (could be genetic, e.g. eye colour, or environmental, e.g. as a result of parenting)

Mutation: mutations arise within living organisms as a result of 'faulty' genetic material (some mutations are harmless, some are beneficial and some are harmful)

Clone: identical copies of living things (occurs naturally and artificially)

2 gene, chromosome, nucleus, cell

3 In classic genetics, the gene is the basic unit of **inheritance**. Genes carry the instructions for different **characteristics**. Genes are, in fact, short lengths of **chromosomes** and long lengths of **DNA**.

4 (a) 46 (as 23 pairs)

5 (c) 23

6 (b) 1 in 2

7 (a) brown eyes; (d) a pre-disposition to diabetes

8 (c) choosing individuals from the same species with desirable features to breed

9

polar bear	to	big feet
camel	to	long eyelashes
seal	to	large amount of subcutaneous fat

Common misconceptions including simple explanations in brackets

Children may think that:

- genes are the sole determinant of the traits in an individual (environmental factors and life style choices – for humans – can affect the traits of an individual);

- living things choose to have particular characteristics in response to their needs, e.g. the thick white coat of a polar bear (plants and animals adapt in response to their environment for survival).

Useful references

Chin, C. and Teou, L-Y. (2010) Formative assessment: using concept cartoon, pupils' drawings, and group discussions to tackle children's ideas about biological inheritance. *Journal of Biological Education*, 44(3): 108–115.

Davies, G. (2005) Stories, fun and games: teaching genetics in primary school. *Journal of Biological Education*, 40(1): 31.

Duncan, R.G., Rogat, A.D. and Yarden, A. (2009) A learning progression for deepening students' understandings of modern genetics across the 5th–10th grades. *Journal of Research in Science Teaching*, 46(6): 655–674.

Chemistry

Materials and their properties including the particulate nature of matter

1 Elements consist of one type of **atom** only. There are 92 naturally occurring **elements** (there are about 109 in total) and these are grouped according to their similar **properties** in the Periodic Table. **Compounds** are formed when two or more substances combine or **bond**

chemically. **Mixtures** are formed when two or more substances are combined physically and the original substances can be **separated** relatively easily.

2   **A** electron; **B** neutron; **C** proton

3   electron   to   a charge of −1 and a mass of 1/2000 amu

    proton   to   a charge of +1 and a mass of 1 amu

    neutron   to   no charge and a mass of 1 amu

4   True

5   (c) 0.0000001 mm

6   **Elements:** hydrogen, copper

    Compounds: pure water, carbon dioxide, sugar

    Mixtures: air, ink, honey

7   oxygen   to   low melting and boiling point   to   covalent (simple molecules)

    copper   to   conducts electricity   to   metallic

    diamond   to   high melting and boiling point   to   covalent (giant structures)

    sodium chloride   to   easily soluble in water   to   ionic

8   Elastic – deforms when a force is applied but returns to its original   – rubber band
                shape when force is removed

    Plastic – is permanently deformed as a result of a force acting on it   – play dough

    Hard   – very difficult to scratch   – diamond

    Tough – does not break or tear easily   – polythene wrapping

    Brittle – breaks easily   – glass

9   **Physical changes:** true – usually reversible, cause changes in the arrangement of the particles in a substance; false – usually irreversible, produce new substances

    Chemical changes: true – usually irreversible, produce new substances; false – usually reversible, cause changes in the arrangement of the particles in a substance

10 solid   to   definite shape, fixed volume, moderate to high density, not compressible

    liquid   to   takes the shape of the container it is in, fixed volume, moderate density, very slightly compressible

    gas   to   no definite shape, no fixed volume, low density, easily compressible

11 **A** melting; **B** solidifying/freezing; **C** evaporating/boiling; **D** condensing; **E** subliming; **F** reverse subliming

12 **True** – the particles move faster, the particles move further apart, the particles have more energy

False – the particles change shape, the particles lose mass, the particles gain mass, the particles are more tightly bonded

13 (a)   energy is transferred to the molecules of the substance

14 **Chemical change:** paper burning, an egg being boiled, grass cuttings decomposing in a compost heap, concrete hardening, an iron nail rusting

Physical change: solid carbon dioxide changing into gaseous carbon dioxide, ice changing to water, a lump of play dough rolled into a 'wiggly worm', sugar dissolving in a cup of coffee, salt added to an icy path

15 **Exothermic:** plaster of Paris and water becomes hotter, nuclear fission

Endothermic: bicarbonate of soda and vinegar becomes cooler, obtaining iron from iron ore

16 (c)   the same mass as the reactants

17

solid in a solid	to	sand and pebbles
gas in a liquid	to	Coca Cola
gas in a gas	to	air
liquid in a gas	to	clouds
solid in a gas	to	smoke
liquid in a liquid	to	milk
solid in a liquid	to	flour in water

18

paper clips and sawdust	to	using a magnet
pebbles and sand	to	sieving
salt and sand	to	dissolving, filtering, evaporating
different coloured inks	to	chromatography
water and alcohol	to	distillation

19 (c)   break up and become so small that they are no longer visible

20 water – solvent; salt – solute; brine – solution

sugar – solute; water – solvent; syrup – solution

alcohol – solvent; plant oils – solute; perfume – solution

pigment – solute; ink – solution; water – solvent

21 (b)   the molecules of the air freshener spread between the molecules of the air

22 (a) A

(b) A polystyrene; B china

(c) Polystyrene is a better heat insulator than china. The tea cools more slowly. Alternatively, you could say that china is a better heat conductor than polystyrene and the tea cools more quickly.

(d) The tea would cool rapidly. Both curves would show a marked drop in temperature.

(e) The tea would cool down but less rapidly than with milk. Both curves would show a marked but gentle drop in temperature.

(f) Heat energy is transferred through the material of both cups by conduction and radiated into the atmosphere around them. The air around both cups is heated and this energy is transferred further into the atmosphere by convection. Some heat energy is transferred to the atmosphere by evaporation.

Common misconceptions including simple explanations in brackets
Children may think that:

- the uses of objects rather than the materials they are made from are used to identify and characterise them (children focus on the object properties rather than the properties of the materials);

- solids are generally large, heavy and inflexible (this makes it difficult for children to understand that sand, salt or flour are solids);

- the terms solid, liquid or gas refer only to ice, water and vapour/steam (the majority of materials can exist in these different physical states);

- melting and dissolving are the same (melting is a change of state from solid to liquid; dissolving occurs when one substance is mixed into a liquid substance to form a solution);

- when a physical change occurs, a new substance forms (physical changes alter the physical properties not the chemical properties of a material);

- substances disappear when they evaporate or dissolve, and can also 'magically' reappear (the notion of conservation of mass is difficult for children to understand).

Useful references
Chambers, B. (1992) A physical presence. *Junior Education*, January, 26–27.

de Boo, M. (2000) Exploring materials in the early years. *Primary Science Review*, 63: 8–10.

Hill, J. (1991) Looking at bricks. *Primary Science Review*, 16: 12–13.

Hope, S. and Cumming, J. (2000) Language and laughter in the kitchen. *Primary Science Review*, 53: 22–23.

Krnel, D., Glažar, S.S. and Watson, R. (2003) The development of the concept of 'matter': a cross-age study of how children classify materials. *Science Education*, 87: 621–639.

Linfield, R.S. (2000) Learning about everyday materials at the science table. *Primary Science Review*, 63: 11–12.

McGuigan, L. (2000) Origins and transformation of materials. *Primary Science Review*, 63: 20–22.

Rees, J. (1998) *That's Chemistry! A resource for primary school teachers about materials and their properties.* London: The Royal Society of Chemistry.

Russell, T., Harlen, W. and Watt, D. (1989) Children's ideas about evaporation. *International Journal of Science Education*, 11(5): 566–576.

Russell, T., Longden, K. and McGuigan, L. (1991) *Materials: Primary SPACE Project Report.* Liverpool: LUP.

Warwick, P. and Stephenson, P. (2000) Heating and changing materials. *Primary Science Review*, 63: 4–7.

Yair, Y. and Yair, Y. (2004) 'Everything comes to an end': an intuitive rule in physics and mathematics. *Science Education*, 88: 594–609.

Earth and atmosphere

1 **igneous** – formed from the intrusion or extrusion and cooling of molten rock – granite, basalt

   metamorphic – formed when heat and pressure completely change existing rocks over long periods of time – slate, schist, gneiss, marble

   sedimentary – formed when layers of sediment get buried and crushed under the weight of other layers – limestone, mudstone, sandstone

2 **A** heat energy from the Sun; **B** evaporation; **C** transpiration; **D** condensation; **E** precipitation; **F** run-off

3 Non-renewable means that the energy resources will one day become depleted and cannot ever be used again or replaced.

4 Renewable means that the energy resources will never run out, can be used over and over, and are replaced continually.

5 (d) the petrified remains of different organisms from a previous geological age

Common misconceptions including simple explanations in brackets
Children may think that:

- any item that is hard is a rock (bricks, hardened clay, concrete, as well as limestone or granite are all thought of as rocks);
- rain occurs when the clouds are full (rain/precipitation occurs when warm moist air meets colder air; the water vapour in the warm air changes from a gas to a liquid and forms rain drops).

Useful references
Blake, A. (2005) Do young children's ideas about the Earth's structure and processes reveal underlying patterns of descriptive and causal understanding in earth science? *Research in Science & Technological Education*, 23(1): 59–74.

Kelemen, D. (1999) Why are rocks pointy? Children's preference for teleological explanations of the natural world. *Developmental Psychology*, 35(6): 1440–1452.

Physics
Motion and forces

1 A force is a **push**, a **pull**, a twist or a turn. When an object is stationary or moving at a constant speed in a straight line the forces acting on it are said to be **balanced**. Unbalanced forces cause objects to start moving and **speed** up, slow down and come to a stop, or change **direction**. Unbalanced forces also bring about changes in **shape**. If an object has no **driving** force moving it along, it will always slow down and stop because of friction. Friction also includes **air** resistance and **water** resistance. Friction always increases as the speed of a moving object increases. Friction can also be useful. Without friction, standing up, riding a bicycle or driving around in cars would be very **difficult**.

2

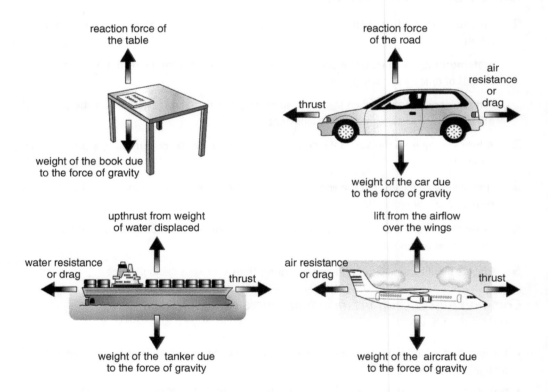

reaction force of the table

reaction force of the road

air resistance or drag

thrust

weight of the book due to the force of gravity

weight of the car due to the force of gravity

upthrust from weight of water displaced

lift from the airflow over the wings

water resistance or drag

thrust

air resistance or drag

thrust

weight of the tanker due to the force of gravity

weight of the aircraft due to the force of gravity

(*Note*: 2 marks for the balanced forces in each example)

3 (c) all of the above

4 (c) all of the above

5 (b) newtons (N)

6 On the surface of the Earth, the hammer would hit the ground first. The feather would be seriously affected by air resistance, which would slow down its fall. On the Moon there is no air. The hammer and the feather would fall together and hit the surface at the same time.

7 Mass and weight are separate things. The mass of an object is simply the amount of **matter** in it. Mass is measured in **kilograms** (kg). The mass of an object remains the same regardless of whether it is found on Earth or in space. Interestingly, any two objects with mass exert a **force** on each other but this is only noticeable when one of the objects is particularly massive. This force of attraction between all masses is called gravity. Weight is a force. As a result, weight is measured in **newtons** (N). The weight of an object changes depending on where it is in the Universe. All objects on the surface of the Earth are pulled towards it with a force of about 10 N/kg. The Moon is much less massive than the Earth. All objects on the surface of the Moon are pulled towards it with a force of about 1.6 N/kg. An object on the surface of the Earth therefore weighs **more** than the same object on the surface of the Moon even though it has exactly the **same** mass.

8 10 x 90 = 900 N

9 1.6 x 90 = 144 N

10 True. An example of gravity forces acting between two objects at a distance is **between** the Earth and the Moon. The Earth's gravitational pull keeps the Moon in orbit although the two objects are not in contact with each other.

11 The movement of an object can be described in terms of its speed or how **fast or quickly** it is going. The speed of an object can be calculated easily if we know the **distance** it travels and the **time** taken to travel that distance. Speed is usually measured in **metres per second** (m/s). The movement of an object can also be described in terms of its velocity. The term velocity should be used in preference to speed when the **direction** in which an object is moving is given. Objects do not always travel at a constant speed or velocity, however. They can always speed up or change direction. Objects which speed up, change direction or do both at the same time are said to **accelerate**.

12 **Speed** = 540/45 = 12 metres per second

Time = 120/12 = 10 seconds

Distance = 15 x 120 = 1800 metres

Sound

1 **A** outer ear; **B** middle ear; **C** inner ear; **D** pinna; **E** ear canal; **F** ear drum; **G** hammer; **H** anvil; **I** stirrup; **J** cochlea; **K** auditory nerve; **L** Eustachian tube; **M** semi-circular canals

Ears allow us to hear. We have two ears in order to help locate sound sources accurately. Sounds entering the ears cause the ear drums to **vibrate**, which in turn force the three small bones of the middle ear to move. Specialised cells within the cochlea (sensitive to vibration and movement) change sound energy to electrical energy. Electrical impulses travel via the auditory nerve to the brain where they are **processed** and interpreted as sound. Unwanted or disagreeable sound is called **noise**. Exposure to particularly loud sounds even for short periods of time can damage the ear drum and lead to partial or complete **deafness**. The ears are never 'switched off'. The Eustachian tube in each ear is responsible for maintaining **pressure** balance between the middle ear and the outside world. The semi-circular canals are concerned with **balance** and orientation and not hearing.

2 (a) about 330 m/s

3 Sounds generally travel **faster** in solids and liquids than in air.

4 Sound travels in waves. These waves are described as **longitudinal** waves.

5 Sound is a form of **energy**. Sounds travel outwards from a **vibrating** source in waves. Sounds will travel in solids, liquids and gases but not in a **vacuum**. Some sounds are louder or quieter than others and differ in **amplitude**. Some sounds are higher or lower in pitch than others and differ in **frequency**. Noise levels are often measured in **decibels** (dB). Sound waves can also be reflected. Sound reflections are referred to as **echoes**.

6 (a) piano – strings struck by key mechanism

(b) drum – skin struck by a drum stick or hand

(c) recorder – air blown into it

(d) guitar – strings plucked or strummed

Pitch can be altered by changing the thickness, the length and the tightness of the strings. Volume can be changed by plucking or strumming with more or less force.

Light

1 **A** eyelid; **B** iris; **C** cornea; **D** pupil; **E** lens; **F** aqueous humour; **G** ciliary muscle; **H** sclera (white of eye); **I** retina; **J** vitreous humour; **K** optic nerve

Eyes allow us to see. Each eye has a series of **muscles** which allow movement within the retaining eye socket. Two eyes provide **binocular** vision, which provides depth perception and distance judgement. Light enters the eye through the cornea and lens which focus the light rays onto a **light**-sensitive layer called the retina. Specialised cells within the retina called **cones** (sensitive to colour) and **rods** (sensitive to 'greys') change light energy to electrical energy. Electrical impulses travel via the optic nerve to the brain where they are **processed** and interpreted as sight. Between the cornea and the lens is the iris. The iris determines the **colour** of the eye. At the centre of the iris is the pupil. The pupil changes size in response to the amount of **light** entering the eye. This means that in areas with **bright** light the pupil will be **smaller** than in darker areas.

2 Short-sightedness or myopia results in light rays being focused 'short' of the retina and within the eyeball itself. This can be corrected using a **diverging** or concave lens. Long-sightedness or hyperopia results in light rays being focused beyond the retina and 'behind' the eyeball itself. This can be corrected using a **converging** or convex lens.

3 **Primary**: a torch, a burning candle, the Sun

Secondary: the Moon

4 (a) 300 000 km/s

5 Light travels in waves. These waves are described as **transverse** waves.

6 Light is a form of **energy**. Light travels in straight lines from a source unless prevented from doing so. Light is a small part of what is referred to as an **electromagnetic** spectrum of waves, which include gamma rays, X-rays, UV, IR, microwaves and radio waves. Light waves have some important features that can be measured: the **wavelength**, which determines the colour of the light, the **frequency**, or the number of waves that pass every second, and the **amplitude**, which determines the intensity or brightness of the light.

7 (b) red, blue and green

8 (c) red, blue and yellow

9 red, orange, yellow, green, blue, indigo and violet

10 (a) Green objects absorb all colour wavelengths except for green.

(b) Objects appear white because they **reflect** all of the light that falls on them. Objects appear black because they **absorb** all of the light that falls on them.

11 Shadows are formed when light is blocked. When the light from a projector is blocked by an object, the shadow formed on a wall, for example, can be made **bigger** by increasing the distance between the object and the screen or by decreasing the distance between the projector and the object. Some shadows are 'black'. Some shadows appear with a dark central area and a fuzzy, grey outline. The dark part of the shadow is known as the **umbra**. The fuzzy, grey outline is known as the **penumbra**.

12

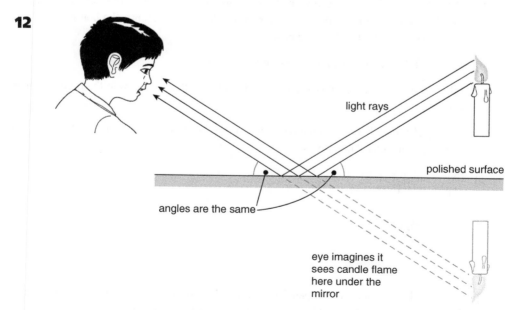

light rays

polished surface

angles are the same

eye imagines it sees candle flame here under the mirror

13 Polished surfaces reflect light perfectly resulting in a clear reflection. Rough surfaces reflect light at different angles resulting in a diffuse reflection.

14 transparent — allows light to pass and objects to be seen clearly

translucent — allows light to pass but objects appear blurred

opaque — blocks light and images completely

Electricity and magnetism

1 (a) True

(b) Unscrewing one bulb from its holder will cause the other bulb to go out.

(c) The same current flows through all parts of a series circuit, so 0.2 A.

(d) The total voltage is shared between components. The bulbs are identical, so 1.5 V.

(e) Rearranging Ohm's Law gives R = V/I, so R = 1.5/0.2 or 7.5 ohms (the total resistance of the whole circuit would be 15 ohms – both bulbs).

(f)

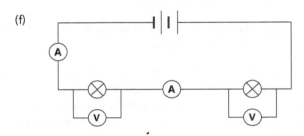

(g) The battery supplies the circuit with electricity and therefore energy. The electrical energy is transferred through the connecting wires to the circuit components, the bulbs, where it is changed or converted to light and heat.

(h) True – the three identical bulbs will be equally bright.

False – the three identical bulbs will be less bright than the two bulbs in the previous series circuit.

Adding a third bulb increases the resistance of the circuit as a whole and thus decreases the intensity of the bulbs' brightness.

(i)

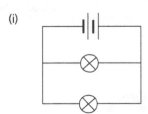

(j) When bulbs are connected in parallel, the overall resistance of the circuit is reduced. Both bulbs would glow with equal brightness but more brightly than in the series circuit.

If one of the bulbs is removed then the other bulb would continue to glow as the circuit is still closed.

2   **A** filament lamp or bulb; **B** cell; **C** battery; **D** switch; **E** resistor; **F** motor; **G** buzzer

3   **Conductor**: a material which allows electricity to flow through it

Insulator: a material which does not allow electricity to flow through it

Current: the flow of electrons or electricity around a circuit

Voltage: the energy of the electrical flow (sometimes referred to as the force that pushes the current around)

Resistance: the difficulty electricity has in passing through a conductor (a resistor is anything that opposes the flow of current)

Power: the rate at which energy is transferred to, say, a circuit component

4   **Good conductors**: copper, aluminium, gold

Poor conductors: skin, water

Insulators: rubber, wood, air

5   Some materials are magnetic. That means that they are both **attracted** to magnets and can be **magnetised**. Magnetic materials include the metals **iron** and **steel (not stainless)** as well as nickel and cobalt. All magnets have **north** and **south** poles. The rule of magnets states that **unlike** poles attract while **like** poles **repel**. That means that when **opposite** poles of magnets are brought close together it is possible to feel an **attractive** force between them. When **like** poles of magnets are brought close together it is possible to feel a **repelling** force between them. A simple compass is nothing more than a freely moving magnet which aligns itself with the Earth's magnetic field. The end of the magnet which points north is referred to as the **north**-seeking pole. The rule of magnets tells us, therefore, that what we refer to as geographical north is actually the Earth's magnetic **south** pole.

6   The magnetic field strength can be changed by altering the size of the current flowing through the wire, the number of winds of wire around the nail, and what material the nail is made of.

The Earth and space

1   The Universe is, quite literally, everything that exists: **matter** (from atoms and molecules to stars and galaxies), **radiation** (visible light together with the rest of the electromagnetic spectrum) and **space** (the vast emptiness within and between galaxies). The Universe is about 12 **billion** years old and most probably emerged from an explosive event referred to as the **Big Bang**. The Universe has been growing in size or **expanding** ever since.

2   Galaxies are assemblages of **stars**, nebulae and other interstellar materials. A typical galaxy contains more than **100** billion stars and measures about **100 000** light years across. Galaxies are classified into four main groups depending on their appearance: **spirals**, barred spirals, **ellipticals** and irregulars. Galaxies are not randomly scattered throughout the Universe; they occur in clusters: **rich** clusters of hundreds or thousands of galaxies and **poor** clusters of a few tens. Our own Sun is located within the Orion Arm of what is referred to as the **Milky Way** galaxy, one of about 30 other galaxies known as the Local Group.

3 Universe, local group of galaxies, Milky Way, Solar System, Earth–Sun–Moon system, Sun, Earth, Moon

4

Sun	to	a star (ball of hot, glowing gas)
planets	to	rocky and gassy objects which orbit the Sun
moons	to	natural satellites which orbit planets
asteroids	to	lumps of rock often referred to as minor planets
comets	to	chunks of ice and other material often seen with a tail
meteoroids	to	small particles of dust and rock fragments

5 (a) Mercury, Venus, Earth, Mars, Jupiter, Saturn, Uranus, Neptune, Pluto

(b) Mercury, Venus, Earth, Mars

(c) Jupiter, Saturn, Uranus, Neptune

(d) False – Venus is, because of its dense atmosphere and greenhouse effect.

(e) Jupiter

(f) Saturn

(g) Mercury and Venus

(h) Jupiter, Saturn, Uranus, Neptune

(i) True

6 **A** where the Sun never 'sets'; **B** where the Sun never 'rises'; **C** where more than 12 hours of daylight are experienced; **D** where exactly 12 hours of daylight are experienced; **E** where less than 12 hours of daylight are experienced; **F** light from the Sun

7 (a) approximately 24 hours

8 The **tilt** of the Earth's axis relative to the plane of its orbit around the Sun causes the seasons. In the UK, the year-long cycle of seasons includes spring, summer, autumn and winter. In June, the **northern** hemisphere is tilted **towards** the Sun and experiences summer while the **southern** hemisphere is tilted away and experiences **winter**. The effects are dramatic. In the UK, for example, summer days are **long**, the Sun 'rises' **high** above the horizon so the Sun's rays reach the surface of the Earth at a **high** angle, and the Earth is **heated** by the Sun for **more** than 12 hours. The Sun's heating effect is **more** efficient and summers are warm. In December the opposite occurs.

9 (c) 365.25 days

10

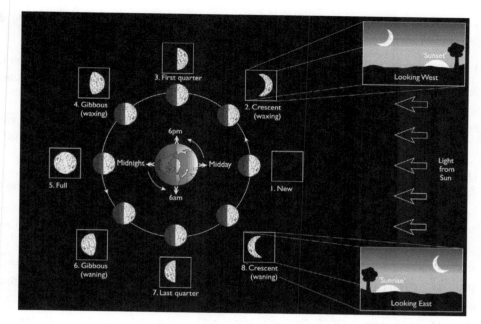

11 (b) once a month

Common misconceptions including simple explanations in brackets
Children may think that:

- when an object is not moving, there are no forces acting on it (objects at rest still have forces such as weight and friction acting on them);

- objects fall 'naturally' (objects fall because of the gravitational pull of the Earth);

- heavy objects fall faster than light objects (all objects fall with the same rate of acceleration; air resistance may slow down the fall of some objects);

- gravity is a property of the Earth only (gravity is a property of all objects with a given mass; the bigger the mass of the object the bigger the gravitational pull it exerts on other objects, e.g. the Earth has a bigger gravitational pull than the Moon);

- all objects that are light float; all objects that are heavy sink (sinking and floating depends on the density of an object; objects that are more dense than water will sink and objects that are less dense than water will float in water – density depends on the volume and mass of an object);

- we see things because there is light coming out of our eyes (we see things because light from a source reflects off objects and into our eyes);

- the Earth is flat (the Earth appears to be flat; models of the Earth can show how it is almost spherical);

- the Moon is a source of light (the Moon is not a source of light as it does not generate light but reflects light off the Sun; the Moon can be thought of as a secondary source of light);

- all metallic objects are magnetic (not all metals are magnetic; metallic objects made from gold, silver and brass are not magnetic but metallic objects from iron and nickel are magnetic);

- a complete circuit consists of one wire with a bulb and a cell at either end of it (for a circuit to be complete, it needs to provide a closed continuous path through which the current can flow through all its components);

- in a simple circuit, electricity moves from both sides of the cell toward the bulb (electric current moves from one terminal of a cell through its components and towards the other cell terminal).

Useful references

Bar, V., Zinn, B., Goldmuntz, R. and Sneider, C. (1994) Children's concepts about weight and free fall. *Science Education*, 78: 149–169.

Bryce, T.G.K. and Blown, E.J. (2013) Children's concepts of the shape and size of the Earth, Sun and Moon. *International Journal of Science Education*, 35(3): 388–446.

Kibble, B. (2011) Day and night: it's obvious how it works, isn't it? *Primary Science*, 116: 18–21.

Sharp, J.G. (1996) Children's astronomical beliefs: a preliminary study of Year 6 children in south-west England. *International Journal of Science Education*, 18(6): 685–712.

Sharp, J.G. and Kuerbis, P. (2006) Children's ideas about the solar system and the chaos in learning science. *Science Education*, 90: 124–147.

Sharp, J.G. and Sharp, J.C. (2007) Beyond shape and gravity: children's ideas about the Earth in space reconsidered. *Research Papers in Education*, 22(3): 363–401.

Siry, C. and Kremer, I. (2011) Children explain the rainbow: using young children's ideas to guide science curricula. *Journal of Science Education and Technology*, 20(5): 643–655.

Vosniadou, S. and Brewer, W. (1992) Mental models of the Earth: a study of conceptual change in childhood. *Cognitive Psychology*, 24: 535–585.

Part 6: Targets for further development

Target setting is an everyday occurrence for most teachers – they do it all the time when assessing and marking children's work and keeping records. Target setting is now seen as a positive step towards helping children make progress. As your own training gets underway, you might well be asked to set targets for yourself. Targets will almost certainly be set for you!

Formally record your targets for further development below. (You might need to copy this sheet and make it larger!) Make **clear** and **specific** reference to areas within your perceived competence and science test that require attention. Don't forget to indicate where, when and how the targets will be achieved.

Targets (areas identified from the audit and test results requiring attention)

Biology	Chemistry	Physics
Structure and function of living organisms – plants:	Materials and their properties:	Motion and forces:
		Sound:
Structure and function of living organisms – animals, including humans:	The particulate nature of matter:	Light:
Interactions and interdependencies:	Earth and atmosphere:	Electricity:
Genetics and evolution:		Magnetism:
		Space physics:

Evidence (record of having achieved targets – nature and location of work undertaken)

Biology	Chemistry	Physics
Structure and function of living organisms – plants:	Materials and their properties:	Motion and forces:
		Sound:
Structure and function of living organisms – animals, including humans:	The particulate nature of matter:	Light:
Interactions and interdependencies:	Earth and atmosphere:	Electricity:
Genetics and evolution:		Magnetism:
		Space physics:

Part 7: Revision and further reading

Well done indeed! Having got this far in the book we can assume that you have managed to work your way through all of the tasks presented. Feeling tired? Take a well-earned rest. As we have emphasised throughout, it really doesn't matter how 'well' or how 'badly' you have done: you have already kick-started the process of learning science and at this stage, that is what matters most. It is our experience that many trainees who recognise that they still have some way to go with science actually make very significant progress with a little bit of hard work and effort. It is also our experience that the vast majority of trainees look upon auditing and testing positively and as a valuable step towards QTS.

Recommending which revision guides to use is always something of a challenge. We have already highlighted a text within the *Learning Matters QTS Series* by Sharp et al. (2012), but there are others too, and we have listed some below. All have been written with primary trainees and primary teachers in mind and all will be found useful in one way or another. Take a look at a few in your local bookstores before buying one. Read it through and then go back to the test for a second time. We think you will be amazed at how much better you will 'perform'.

Further reading

Allen, M. (2010) *Misconceptions in Primary Science*. Maidenhead: Open University Press.

Farrow, S. (2006) *The Really Useful Science Book*. 3rd ed. London: Falmer.

Harlen, W. and Qualter, A. (2009) *The Teaching of Science in Primary Schools*. 5th ed. London: David Fulton.

Hollins, M. and Whitby, V. (2001) *Progression in Primary Science: A Guide to the Nature and Practice of Primary Science at Key Stages 1 and 2*. London: Fulton.

Howe, A., Davies, D., McMahon, K., Towler, L., Collier, C. and Scott, T. (2009) *Science 5–11: A Guide for Teachers*. 2nd ed. London: Routledge.

Peacock, G. (2002) *Teaching Science in Primary Schools: A Handbook of Lesson Plans, Knowledge and Teaching Methods*. 2nd ed. London: Letts.

Peacock, G., Sharp, J., Johnsey, R. and Wright, D. (2012) *Primary Science: Knowledge and Understanding*. 6th ed. London: Learning Matters/SAGE.

Roden, J., Ritchie, H. and Ward, H. (2007) *Primary Science: Extending Knowledge in Practice*. Exeter: Learning Matters.

Rutledge, N. (2010) *Primary Science: Teaching the Tricky Bits*. Maidenhead: Open University Press.

Wenham, M. and Ovens, P. (2010). *Understanding Primary Science: Science Knowledge for Teaching*. 3rd ed. London: SAGE.

Online reference sources

Association for Science Education: www.ase.org.uk

BBC learning home page: www.bbc.co.uk/learning

BBC schools home page: www.bbc.co.uk/schools

Science Museum: http://sciencemuseum.org.uk/Home/educators.aspx

The Natural History Museum: www.nhm.ac.uk/education/index.html

Exploratorium: www.exploratorium.edu/education

Nasa Kids' Club: www.nasa.gov/audience/forkids/kidsclub/flash/index.html

Keywords

primary science; elementary science; inquiry; enquiry; practical work; practicals; hands-on science; lab work; science investigations; children's ideas in science, misconceptions in science; alternative conceptions in science

(Note: Website addresses are prone to regular change. Use key words such as the ones suggested above in any search engine to track down the sites above and other online resources for primary science.)

Achieving QTS

Primary Science: Audit and Test

First published by Parragon in 2012
Parragon
Queen Street House
4 Queen Street
Bath BA1 1HE, UK
www.parragon.com

That's When I'm Happy: Text © Beth Shoshan, Illustration © Jacqueline East. First published 2005 by Meadowside Children's Books.
If You Can...We Can: Text © Beth Shoshan, Illustration © Petra Brown. First published 2008 by Meadowside Children's Books.
Little Rabbit Waits for the Moon: Text © Beth Shoshan, Illustration © Stephanie Peel. First Published 2004 by Meadowside Children's Books.
Cuddle!: Text © Beth Shoshan, Illustrations © Jacqueline East. First Published 2006 by Meadowside Children's Books.
Wide Awake Jake: Text © Rachel Elliot, Illustration © Karen Sapp. First published 2004 by Meadowside Children's Books.
If Big Can...I Can: Text © Beth Shoshan, Illustration © Petra Brown. First published 2006 by Meadowside Children's Books.
My Favorite Food: Text © Tiziana Bendall-Brunello, Illustration © John Bendall-Brunello. First published 2010 by Gullane Children's Books.

Published by arrangement with Meadowside Children's Books and Gullane Children's Books
185 Fleet Street London EC4A 2HS

ISBN 978-1-4454-8921-6

Printed in China

Goodnight
Stories Collection

PaRRagon

Bath • New York • Singapore • Hong Kong • Cologne • Delhi
Melbourne • Amsterdam • Johannesburg • Auckland • Shenzhen

Contents

Cuddle!

I'd cuddle a whale,
but I might be too small,

I'd cuddle
a hedgehog
but, **ouch!**
they're so spiky,

11

I'd cuddle a crocodile.

Hmmm…?

Not likely!

13

If I cuddled
a gorilla

I would end up
much thinner,

If I cuddled a tiger
I'd end up as dinner.

I'd cuddle a skunk
but I think they're
too smelly,

I'd cuddle a shark

but I'd be in his belly!

21

I'd cuddle
a python

way up high
in a tree,

23

I'd cuddle a hippo
who might just
squash me.

24

I'd cuddle a lion

but he'd bite off my head,

25

Do you think
I can cuddle
**my teddy bear
instead?**

Little Rabbit *Waits for the* Moon

Little Rabbit
couldn't sleep...

29

In the day,
the sun is there, warm
and bright. But when night comes,
the sky hangs low, dark, and empty.

30

"If I fall asleep now, there'll be no one watching over me," thought Little Rabbit. "I'll just have to wait for the moon." And so he did just that.

The trouble with being so tired and sleepy,
was that he didn't know exactly when
the moon would come.

More time passed and the moon still hadn't come.

"This is my first day, ever,"
said a small flower in the fields.
"Maybe I will have grown into a tree
by the time your moon comes."

That sounded like a very long time.

Little Rabbit thought he had better ask
someone else—just to be sure.

35

"Look deep into the water," shimmered a little lake nearby. "Maybe your moon has fallen in and can't get out."

That didn't sound like what he wanted to hear.

Little Rabbit thought he had better ask someone else—just to be sure.

"Why don't you walk with me?" twisted a long and winding path. "We can find out where I'm leading and maybe your moon is at the other end!"

That sounded like it might be a long way away.

Little Rabbit thought he had better ask someone else—just to be sure.

"I've just blown in to these parts," breezed a wind that had picked up. "Who knows? I might be a big, fierce storm by the time your moon comes."

That didn't sound like something he wanted to wait for.

Little Rabbit thought he had better ask someone else—just to be sure.

41

42

"We can't see your moon yet," rumbled the great, rolling hills. "And we can see far into the distance from up here!"

That didn't sound very promising.

Little Rabbit began to think that the moon might never come. And he was getting very, very tired…

And then, from behind the hills,
carried by the wind along the twists
of the path, reflected in the lake, and shining
on the petals of the small flower...

44

...the most perfect moon
slid into the night sky.

But Little Rabbit
had fallen asleep, dreaming
of the moon that would
watch over him through the night.

If Big can...
I can...

If Big can run...
...then I can run

(Though not as fast,
I'm only small)

48

If Big can jump…
…then I can jump

(And Big leaps long,
long, long away)

If Big can swing…
…then I can swing

52

(As Big swings
high into the sky,
I'll get there too,
one day, I'm sure)

53

If Big can climb…
…then I can climb

(It's just I have to take
my time to get to all the
places Big can climb to
with one stretch)

55

If Big can see...
...then I can saw

56

(Up, up I go, into the air.
I never thought I'd get so high,
but now that Big is on the ground
how will I get down?)

57

If Big can play…

...then I can play

(Deep in the sand Big digs a hole...

...but if I fell in I'd be stuck and then I'd have to yell and shout for Big to come and pull me out)

But what if Big
can't get in places
only I can squeeze inside?

61

I'd find the treasures,

Rule the roost!

Be number one...

And Big would know
I'm having fun.

If Big could only see…

...how great it is
inside my den and all
the games I like to play
on oceans around the world
and spaceships in the sky.

Here I can
swing into
the air!

Running fast
and leaping long

and stretch to climb
and soaring high
into the air
and digging deep...

...but Big's not there...

67

...I'm all alone...

...and that's no fun,

so...

Whatever Big can...

70

and whatever I can...

We can…

...together!

That's When I'm Happy!

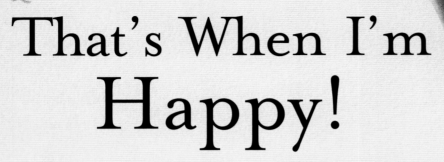

There are some days
when I'm very happy...
and there are some days
when I'm a little bit sad.

But now, on those days
when I'm a little bit sad...
I try and find my way back
to being happy.

75

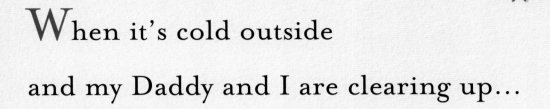

When it's cold outside
and my Daddy and I are clearing up...

And when we take a soft striped bag
and fill it up with leaves...

And when he chooses
one special leaf for me
because it's deeper, darker,
redder than all the others...

76

And when my Daddy
and I play soccer
through the leaves
and we're kicking
to each other
through a tunnel of trees...

...that's when I'm happy!

When it's cozy inside and my
Mommy and I give each other
great big bear hugs…

And when we rub our noses together…

And when she chooses one
special tickle, just for me
because it's wiggly,
squirmy, and makes me laugh
more than all the others…

82

And then my Mommy
gives me the biggest kiss of all…

And I reach up to give her a big kiss back…

But not as big,
because my mouth is still very small…

…that's when I'm happy!

83

When it's night outside

and my Daddy and I gaze through the window…

And when he takes my hand

and points at the night sky…

And when he chooses one special star

for me because it's bigger,

burning brighter than all the others…

And then my Daddy
and I count all the stars
in the sky, and he says
there are more than 119...

But I can't count any higher...

...that's when I'm happy!

87

When it's warm inside and my Mommy

and I run our fingers through the books…

And when we look at all the pictures…

And when she chooses

one special book for me

because it's our favorite,

better than all the others…

89

And then my Mommy reads
the perfect story to me
and I can read some
of the words...

But mostly the ones
with the letters
from my name in them...

...that's when I'm happy!

When it's dark everywhere
and I cuddle up
to my Mommy and Daddy
(even though they're asleep)
still telling stories to myself,
watching stars in the sky,
bathed in all their kisses,
and dreaming of the deep
red leaves…

That's when we're happy!

If You Can… We Can!

I love you…

I really do!

(Although my arms are just too small
and so I can't quite cuddle you.)

I hug you...
you hug me.

(And around
and around
we dance together,
holding tight.

Don't let me fall!)

99

100

I tickle you...
you giggle too.

(But not my toes...! No!
Not my toes,
you know that's when I'll squeal the most!)

I make you laugh...
you laugh with me.

(There's nothing in this world
that can make us feel so good
as laughter can, as laughter does,
as laughter should.)

I hold your hand…
you hold mine tight.

(Just feeling snug, secure, and safe.
Just knowing you'll protect me,
care for me…
be there.)

I sing you songs…
you sing them too.

(Loud ones, soft ones,
make-me-laugh ones.
Love songs, sleep songs,
safe-and-sound songs.)

I tell you tales...
you listen close.

(Then tell me stories
through the night...

of mighty dragons,
gallant knights...

adventures made
to fill my mind.)

I'm in your dreams…
and you're
in mine.

(The best dreams, safe dreams,
sleep-all-night dreams.
My dreams, your dreams.
Always our dreams.)

Let's be friends forever, I say!

There for one another,
looking out and taking care.

So…

112

Whatever you do...

and whatever I do...

Let's do it...

...together!

Wide Awake Jake

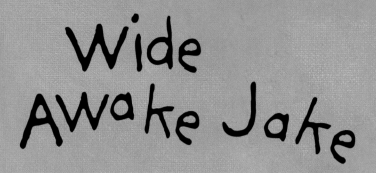

Jake couldn't sleep.

He lay on his back.

He lay on his tummy.

He even lay upside down.

But it was **no** good.

He was **still** wide awake.

So Jake got up and went downstairs...

HUFF

PUFF

HUFFITY

PUFF

120

"I can't sleep," huffed Jake.

"Try counting sheep," said Dad.

But Jake didn't think **that** would help.

"Pretend you're a little bear,
going to sleep for the winter,"
said Mom.

Jake thought that **could** work.

So he went back upstairs...

pad

pad

pad

Jake curled up inside his blanket.

"I'm a little bear,

grrr, grrr,"

he growled.

But then he heard noises.

What if it was a great BIG bear?
It might have long, sharp claws
and huge, yellow teeth!

And his fur was very itchy.

124

Jake, the little bear,
was still wide awake.
So he got up

and went downstairs...

125

THUMP
BUMP
CLUMPITY
THUMP

126

"I can't sleep," grumbled Jake.

"Count to a million," said Dad.

Jake didn't think **that** would help.

"Pretend you're a little mouse,
going to sleep in a mousehole,"
said Mom.

Jake thought that **could** work.

So he went back upstairs...

squeak

squeak

squeak

Jake crawled to the bottom of his bed.

But then he heard noises.

It might be a
big
fat
cat!!!

"Eek!" squeaked Jake,
the little mouse.
HURRY
FLURRY
SCURRY

"I can't sleep," worried Jake.

Dad just sighed.

Jake didn't think **that** was very helpful.

"Pretend you're a baby bird
in your nest,"
said Mom.

Jake thought it was worth a try.

So he went back upstairs...

flutter

flutter

flutter

Jake pulled his pillow under the covers.

"I'm a baby bird,
sitting in my nest,"
he whispered.

But his feathers kept

making him sneeze.

Then he
heard
noises!

Somebody was
pulling the
covers
down...

It was a big, hairy bear!

No, it didn't have any claws.

It was a fat, scary cat!

No, it didn't have a tail.

It was a big, scowly owl!

No, it didn't have a beak.

It was Mom!

Mom put the pillows straight.
She tucked Jake in,
nice and **tight.**

"You're my **brave** little Jake,
safe in your very own bed,"
said Mom.

"Now close your eyes.
It's time to sleep."

And with a **growly** yawn,

and a **mouse-quiet** wink,

and a fluttery blink,

Jake, the little boy,
was fast asleep.

139

My Favorite Food

Little Goose and her mommy were in the yard, enjoying some fresh, green grass.

"Mmm . . . I love grass,"
said Little Goose. "It's
my favorite food!

I wonder if everybody
loves grass as much as me?"

"Why don't you go and
find out," said Mommy.

So off went Little Goose to
find out Pig's favorite food . . .

"What's your favorite food, Pig?"
she asked.

"Apples,"
said Pig. "They're so juicy!"

142

"Mmm," said Little Goose, "I like apples too.
I wonder what Goat's favorite food is?"

So off she went to find out . . .

145

"What's your favorite food, Goat?"
asked Little Goose.

"Socks,"
said Goat. "They're so chewy!"

146

147

"Hmmm," said Little Goose,
"I'm not sure I like socks!

I wonder what Cow's favorite food is?"

148

So off she went to find out .

149

"What's your favorite food, Cow?" asked Little Goose.

"Daisies," said Cow. "They're so sweet!"

151

"Mmm," said Little Goose, "daisies are tasty.
But I wonder what Fox's
favorite food is?"

So off she went to find out . . .

153

"Fox! Fox! What's your favorite food?" asked Little Goose.

"Well . . ." said Fox,
"let me think.
My favorite food is . . ."

155

"YOU!"

"Yikes!" squealed Little Goose.
And she ran away as fast as her
little legs would carry her . . .

157

. . . safely back into the loving wings of her mommy.

And while Little Goose enjoyed some of her
favorite food—grass—

Fox settled down to eat his favorite food—strawberries!